A series of student texts in

CONTEMPORARY BIOLOGY

General Editors:

Professor E. J. W. Barrington, F.R.S.
Professor Arthur J. Willis

The Biology of Lichens

Second Edition

Mason E. Hale, Jr.

B.S., M.A., PH.D.

Curator, Department of Botany, Smithsonian Institution

American Elsevier Publishing Company, Inc.
New York

American Elsevier Publishing Company, Inc.
52 Vanderbilt Avenue, New York, N.Y. 10017

First published in Great Britain by
Edward Arnold (Publishers) Ltd.

Cloth edition ISBN: 0-444-19531-9
Paper edition ISBN: 0-444-19530-0

Library of Congress Catalog Card Number: 74-12607

Printed in Great Britain by
William Clowes & Sons, Limited
London, Beccles and Colchester

Preface to the Second Edition

Lichenology is the study of a unique group of plants that consist of two unrelated components, fungi and algae, living in a close symbiotic association. They are often given only passing reference in textbooks, claimed by neither mycologists nor algologists. Though sometimes included with mosses, because they occupy similar habitats in nature and are curated by similar methods, lichens and mosses are as far apart as moulds and seaweeds.

The study of lichens has made and will continue to make significant contributions to biology, in spite of the relative lack of economic importance of these plants. As the most perfect examples of two organisms living together in mutual dependence, lichens have shed much light on the broad problems of symbiosis. Because of their physiological sensitivity, lichens are reliable indicators and useful biological measuring devices for atmospheric pollution. They have gained notoriety as accumulators of radioactivity that is passed on to humans through grazing animals. In a more academic area, lichens have set a standard in the field of biochemical systematics, the use of chemistry in taxonomy.

Although lichens hold a continuing fascination for botanists, there are few textbooks available in this field. Advanced texts in French by des Abbayes,[1] Moreau,[200] and Ozenda[209] and a popular account in German by Follmann[94] are of limited use to English-speaking students. The author's[120] semi-popular *Lichen Handbook* is aimed at American lichenologists. Annie Lorraine Smith's *Lichens*[248] truly sets a standard for scholarship, although it is now out of date in many aspects. Recent review articles by Ahmadjian,[8] Haynes,[132] Llano[183] and Smith[250] are useful but probably too specialized for general purposes and lack illustrations.

Since its appearance in 1967, *The Biology of Lichens* seems to have satisfied the need for more detailed information than is available in semi-popular texts, yet which does not approach the encyclopaedic realms of the professional's specialist literature. This new edition updates the discussions on such subjects as lichen chemistry, and effects of pollution on lichens. New materials have been added to cover the rapidly growing field of ultra-structural research; recent work with both the electron and scanning-electron microscopes is reviewed. Part of the chapter on ecology has been rewritten to reflect the European methods of lichen phytosociology more fully. Finally, a number of illustrations have been replaced and over 50 new references added.

It is hoped that these changes will enhance the value of *The Biology of Lichens* as a succinct but readable review of the whole field of lichenology.

Washington   M.E.H.
1974

Table of Contents

I

Morphology of the Thallus

The study of lichen morphology began with the work of Erik Acharius, a Swedish doctor who is regarded as the father of lichenology. In 1803 he introduced the terms soredia, isidia and cephalodia to describe unique lichen structures, and by 1866 the renowned German mycologist Anton de Bary [39] was able to write an accurate account of the morphology of lichens that can still be used as a reference. Annie Lorraine Smith's *Lichens*[248] has an exhaustive summary of descriptive morphology which should be consulted by serious students; Ozenda's[209] well-illustrated text in French brings this field up to date.

STRUCTURE OF THE THALLUS

The lichen thallus is a vegetative plant body of remarkable complexity having little resemblance externally to either non-lichenized fungi or algae. The fungal component (mycobiont) is an Ascomycete or Basidiomycete that has succeeded in establishing a symbiotic relationship with algae. The algal component (phycobiont) is essentially the same as various types of free-living algae. By the symbiotic union of these diverse components, the thallus arising forms an autonomous organism that is in the minds of many botanists a distinct kind of plant, notwithstanding its dual nature. When the thallus is sectioned and examined under a microscope, however, the fungal and algal elements are easily distinguished (Plate 2A).

The basic building blocks of fungi are elongate cellular threads called hyphae. Masses of hyphae form a vegetative thallus or mycelium, which is usually inconspicuous in non-lichenized fungi but often quite elaborate

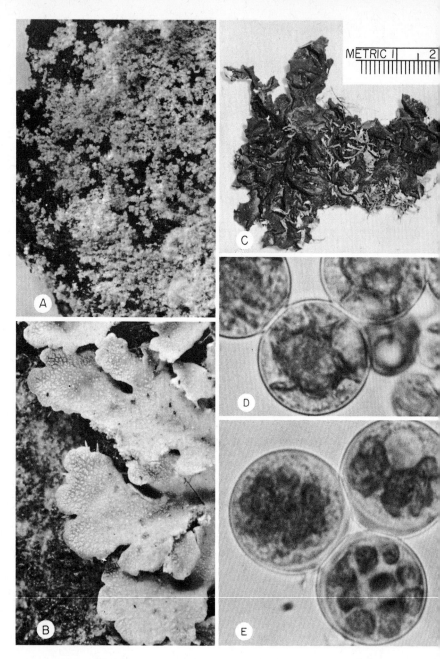

Plate 1 (A) Thallus of *Lepraria incana* (× 12). (B) Pruina on the upper surface of *Physcia grisea* (× 12). (C) Thallus of the gelatinous lichen *Collema subfurvum*. (D) Cells of *Trebouxia* from *Parmelia caperata* (about × 1200). (E) Aplanospore formation in *Trebouxia* from *Candelaria concolor* var. *antarctica* (about × 1200) (D and E by V. Ahmadjian)

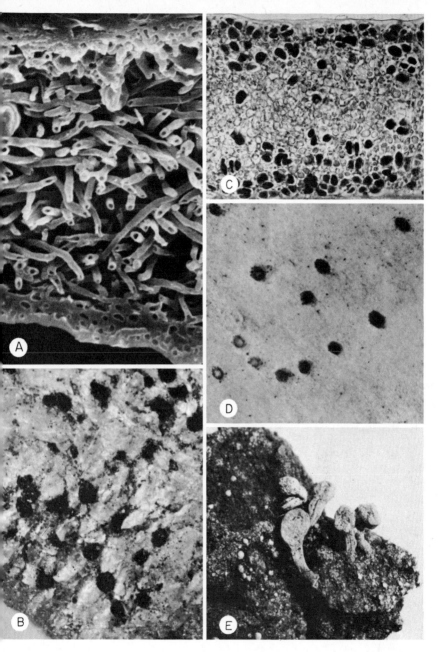

Plate 2 (A) Cross-section of foliose *Parmelia xanthina* (× 800). (B) Endolithic thallus of *Lecidea liguriensis* (× 12). (C) Longitudinal section of the thallus of *Ephebe brasiliensis* (× 120). (D) Pseudothecia of the hypophloeadal lichen *Leptorhaphis epidermidis* (× 12). (E) Stiped apothecia of *Baeomyces roseus* (× 5). (C by A. Henssen)

and durable in lichens. The simplest lichen thallus is a scurfy powdery crust of loosely associated fungi and algae, as in the genus *Lepraria* (Plate 1A), which grows on soil, rock and tree bark. In more highly structured lichens the hyphae predominate in bulk and assume a more regular arrangement, as specialized tissues, a discrete thallus and finally fruiting bodies take shape. The algae are usually limited to a thin layer just below the surface of the thallus. There is a division of labour between the components and a morphological and physiological balance is reached.

Tissues

Cortical layers

The cortical layers serve as a protective covering over the thallus comparable to the epidermis of a green leaf. They are composed of more or less compressed, heavily gelatinized hyphae firmly cemented together. When cellular structure is discernible, as in *Sticta, Peltigera* or *Physcia* (Fig. 1.1a), the cortex is called pseudoparenchymatous or paraplectenchymatous, in

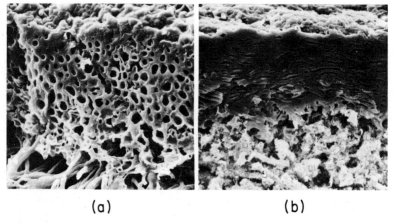

(a)   (b)

Fig. 1.1 Cortical structure of lichens: (a) paraplectenchymatous cortex of *Physcia ciliata* (× 450, long. section); (b) prosoplectenchymatous cortex of *Anaptychia hypoleuca* (× 400, long. section). (SEM photographs)

contrast to the true cellular parenchymatous cortex of higher green plants. When the cells are oriented parallel and the walls are indistinct, the term prosoplectenchymatous is applied. This type is characteristic of *Anaptychia* (Fig. 1.1b) and the lower cortex of many *Physciae*. Many crustose lichens have only a very weakly structured upper cortex. Cortical hyphae are in part perpendicular to the surface in *Roccella, Pseudevernia* and in the rim of the fruiting bodies of some crustose lichens (see Fig. 2.5).[127E]

The upper cortex is usually 10–15 μm thick with several layers of cells, although in some of the gelatinous lichens (Lichinaceae and Collemataceae) a cortex is lacking or only one or a few cells thick (Plate 2C). The surface of the cortex is often covered with a very thin homogeneous cuticle, while a number of lichens, especially in the family Physciaceae, have a whitish surface pruina that may have a morphological origin as necrotic hyphal layers or may be an accumulation of carbonates and oxalates (Plate 1B). Certain orange and yellow pigments, including anthraquinones, pulvic acid derivatives and usnic acid, are usually deposited only in the cortex. The lower cortex, normally present in foliose lichens, is similar to the upper cortex in thickness and structure. While bare in a few genera, the lower surface may be invested with rhizines (Fig. 1.2) or other organs for attachment.

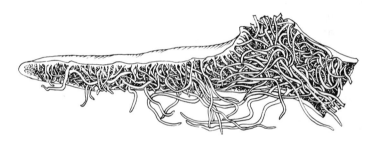

Fig. 1.2 Rhizines on the lower surface of a lobe of *Parmelia fistulata* (× 8). (Drawing by L. Anderson)

Medulla

The bulk of a lichen thallus consists of medullary tissue which may be as much as 500 μm thick. The hyphae are in general only weakly gelatinized and have relatively large cell lumina. There is little compression; the hyphal threads are irregularly interwoven in a loose, often fibrous or cottony layer (Plate 2A). Oil cells intercalated along the hyphae are especially common in endolithic lichens. The medulla has a greater water-holding capacity than any of the other tissues and is a region of food storage. Most colourless lichen acids are deposited here, but it has not yet been determined if they are synthesized *in situ* or in the algal layer and then transported to the medulla.

The medulla of crustose lichens attaches the thallus directly to the bark or rock substrate, some of the hyphae being rhizoidal. If the thallus becomes elevated above the substrate, as in squamulose lichens, no protective layer forms. In assuming the functions of a cortex, the medulla shows no particular specialization except in a few genera such as *Anaptychia* where some species produce conspicuous orange pigments on the exposed hyphae.

Algal layer

Lichen algae are completely surrounded by fungal tissue in the lichen thallus. They are ordinarily confined to a distinct layer about 10–15 μm thick between the upper cortex and the medulla, an arrangement that is called stratified or heteromerous (Plate 2A). The foliose gelatinous lichens, especially those with *Nostoc*, which have poor internal differentiation between the algal layer and medulla, are considered to have an unstratified or homoiomerous type of arrangement. These should not be confused with the gelatinous fungi (order Tremellales) which are not associated with algae. Most filamentous lichens also lack a clear line of demarcation between the algae and hyphae (Plate 2C).

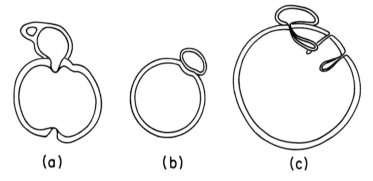

(a)   (b)   (c)

Fig. 1.3 Haustoria invading algal cells : (a) *Cladonia uncialis* ; (b) *Parmelia conspersa* ; (c) *Lecanora varia* (magnified). (After Plessl[214])

The algal cells and the fungal hyphae are in close physical association within the lichen thallus, but the exact relationship between haustoria (special food-absorbing hyphae) and the algal cell wall and protoplast is still to be precisely determined. After a survey of 95 lichens with a light microscope, Plessl[214] found not only evidence of haustorial connections in all but three species but a rather specific correlation between the type of haustoria and the growth form of the thallus. Most foliose and fruticose lichens had intramembranous or appressorial haustoria that clasp algal cells but do not penetrate the living cell (Fig. 1.3a,b). Crustose lichens appeared to have intercellular haustoria that actually penetrate the protoplast (Fig. 1.3c). Moore[199] has shown with electron microscopy that haustoria occur in *Cladonia cristatella* and that these are probably intramembranous.

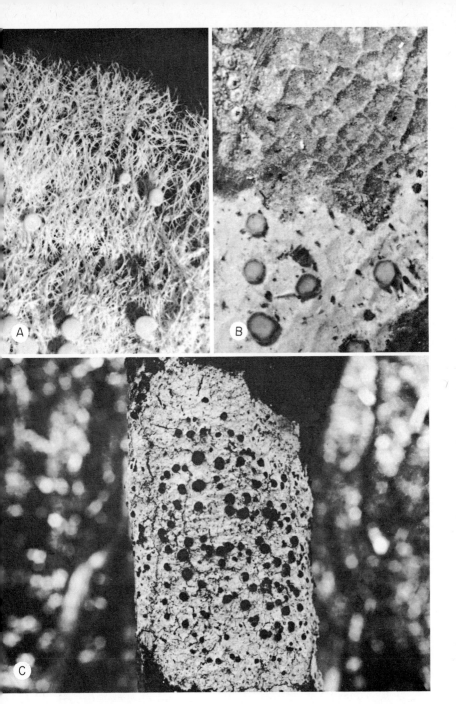

Plate 3 (A) Filamentous thallus of a species of *Coenogonium* (× 5). (B) *Bacidia elegans*, a foliicolous lichen, on an evergreen angiosperm from Sabah (N. Borneo) (× 12). (C) A species of *Catillaria* in Hawaii (× 1/2).

The shape and form of a mature lichen thallus are determined almost exclusively by the fungus, and the specific characteristics of the alga are obscured. Even among the filamentous lichens, the Lichinaceae and Coenogoniaceae (Plate 3A), in which the algal component makes up the mass of the thallus, the fungus has the greater influence, since it is able to combine with different phycobionts and still retain the same thallus morphology.[271] To a certain extent, *Collema*, a gelatinous lichen (Plate 1C), is equally the product of the fungus and *Nostoc*,[78] which, cultured alone, forms colonies similar to the composite thallus. The phycobiont is responsible for the colour, overall shape, and gelatinous consistency of the thallus, whereas the fungus moulds the specific details of lobation and isidia.

IDENTIFICATION OF THE PHYCOBIONT   Specific identification of algae in the thallus is often difficult, sometimes impossible, because the algal colonies have been modified by the fungus to the point of being unrecognizable. An added handicap is that the phycobionts of relatively few lichens have been positively identified in the light of modern algal taxonomy, and many old reports, especially those of Zahlbruckner which are often quoted,[285] are probably erroneous and must be re-examined. Ahmadjian[2] has constructed useful keys for the identification of lichen algae. Ideally the algae should be isolated and subcultured so that formation of zoospores and other reproductive structures can be observed, but this is possible only with fresh material.

The commonest alga in lichens is the green unicell *Trebouxia* (Plate 1D), a close relative of the free-living genera *Cystococcus* and *Chlorococcum*. Santesson[232] estimated that 83% of the lichens in Scandinavia contain this alga. The green filamentous genus *Trentepohlia* is a common phycobiont in crustose forms (*Graphis*, *Arthonia*, *Opegrapha*) but the filamental structure is highly modified. The ubiquitous unicell *Pleurococcus* is also reported as a phycobiont in several crustose genera but on the whole is comparatively rare in lichens. Various blue-green algae, *Nostoc*, *Scytonema*, *Stigonema* (Fig. 1.4), *Hyphomorpha*, and *Calothrix*, are the algal components of the dark brown or black gelatinous and filamentous lichens as well as some or all of the species in the families Coccocarpiaceae, Pannariaceae, Peltigeraceae and Stictaceae, which have stratified thalli. Altogether, some 32 genera, 21 green and 11 blue-green, have been identified as the phycobionts of lichens and more will undoubtedly be found.[2]

SPECIFICITY OF LICHEN ALGAE   There is no evidence that each lichen species has a specific phycobiont or that new races of algae are created by lichenization. Quite the contrary, many different lichen species and even genera have identical algae, and the same lichen can have different algae, though this is a rare occurrence.[3] Most lichen thalli contain only one kind of alga, but the strange subfruticose genus *Compsocladium* from New

Guinea has a blue-green, *Scytonema*, in the lower branches, and a *Chlorella*-like green unicell in the upper branches.[168] Many different species and physiological races of phycobionts, in particular of *Trebouxia*, have been described from lichens. Some of these are based on differences in peptone or amino acid utilization, and in respect of the effects of temperature, pH

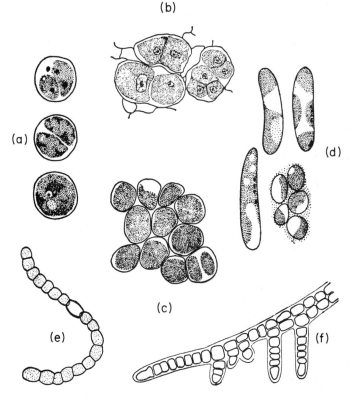

Fig. 1.4 Some algae of lichens : (a) *Trebouxia* ; (b) *Coccobotrys* ; (c) *Chlorella* ; (d) *Coccomyxa* ; (e) *Nostoc* ; (f) *Stigonema* (magnified). (a, c and d from Zehnder,[287] b from Stevens,[256] e and f from Černohorský *et al.*[60])

and other responses in culture; some are based on morphology of the pyrenoid or chloroplast or method of zoospore formation. There is an unfortunate tendency to describe phycobionts from a single lichen, but such a narrow concept implies a much higher degree of algal specificity than is now suspected. In any event, our knowledge of the speciation and constancy of phycobionts is extremely limited and many thousands of lichens remain to be studied.

REPRODUCTION   Reproduction of the algae within the lichen thallus occurs by mitotic cell division and formation of akinetes and aplanospores (Plate 1E). Paulson[211] observed the abundant formation of 'zoogonidia' (aplanospores derived from arrested zoospores) in *Evernia prunastri*, which contains *Trebouxia*, and presumably this type of reproduction is normal in lichens with trebouxoid algae. The parent cell divides into 2 to 16 separate protoplasts which secrete cell walls and burst out, leaving behind the empty mother cell. Zoosporangia, motile zoospores and isogametes are rarely seen in the composite thallus although they can be induced to form when the phycobiont is isolated in pure culture.[2]

ULTRASTRUCTURE   A number of recent studies have been concerned with the internal tissues and cell structure of lichens as viewed with the electron microscope.[99A, 149A] One of the most interesting discoveries is the presence of electron-dense lipid-containing globuli associated with the pyrenoids in *Trebouxia* but unknown so far in free-living algae which usually store starch in plates around the pyrenoid. This new kind of storage product may have evolved as a result of the symbiotic association. In addition, the mycobiont has unique ellipsoidal bodies of unknown function. The hyphal cell walls are surrounded by a thick fibrillar polysaccharide material, infiltrated with bacteria. Much work remains to be done, however, before the ultrastructure of lichen cells is fully understood.

The scanning-electron microscope came into general use with lichens in 1968[212A] and has greatly improved our knowledge of the surface structures of lichens.[131A] Many species of the foliose genus *Parmelia*, for example, are now known to have an epicortical layer covering the cortex and perforated with holes, an obvious aid in gas exchange (Fig. 1.5).[127E]

GROWTH FORMS

Lichens are traditionally classified into three growth forms: crustose, foliose and fruticose. These forms are no more than points on a scale of continuous differentiation from primitive to highly structured thalli but there are of course many intermediates between the three classical types. Each growth form is characterized by a particular arrangement of cortical, algal and medullary tissues and by different degrees of attachment to the substrate.

Crustose lichens

Crustose lichens (Plates 3, 4, 9) are usually in such intimate contact with the substrate that they can scarcely be separated from it. In an extreme example, such as the endolithic species of *Lecidea* or *Sarcogyne*, the upper cortex is entirely absent, the algae are scattered below the outermost rock crystals, and the medullary hyphae penetrate irregularly several millimetres into the rock. Only the ascocarps at the rock surface give any clue to the presence of a lichen (Plate 2B). Hypophloeadal lichens on trees often have a similar relation of the algae and medulla to the cork cells of the bark and they too are visible only by their ascocarps (Plate 2D).

Increasing specialization of the crustose thallus is accompanied by the formation of a distinct upper cortex, algal layer and medulla. The border of the thallus may be diffuse and not sharply delimited from the surrounding substrate (Plate 2D). In many species there is a clearly defined black non-lichenized hypothallus (Plate 4A) which precedes the margin of the lichen-ized thallus. The conspicuous white border of *Pertusaria*, *Catillaria* and some other crustose genera may also be non-lichenized for a considerable distance (Plate 4C). Foliicolous lichens are typically crustose and occur on the upper, rarely lower, surface of leaves of evergreen shrubs, ferns or trees in tropical areas (Plate 3B).[232]

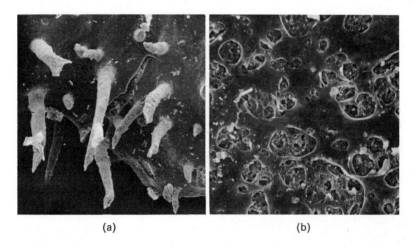

(a)   (b)

Fig. 1.5 Scanning-electron photographs of lichens: (a) rhizines of *Parmelia saxatilis* (× 50); (b) epicortical pores on the surface of *P. gigas* (× 400).

Further development of the crustose thallus can be followed in the genera *Acarospora* and *Rhizocarpon* which have a centrally fissured or areolate thallus (Plate 4A, B). More highly developed forms, such as the marginally lobate species of *Rinodina* (Plate 4D) and *Lecanora*, have distinct lobes differing from those of a foliose thallus in lacking a lower cortex. The ultimate stage of specialization is the squamulose thallus, a discrete lobe-like structure partially or wholly free of the substrate. The medulla is

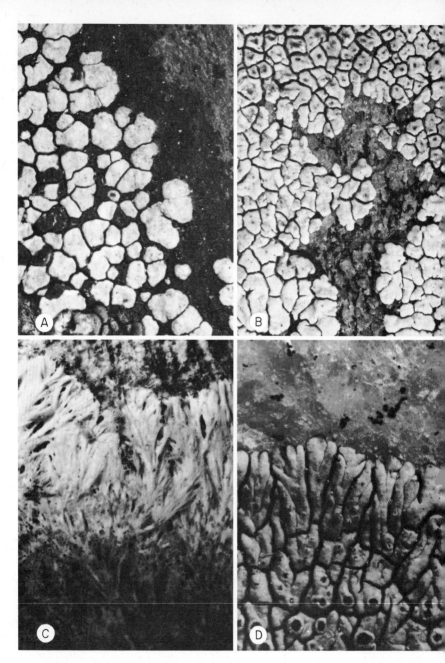

Plate 4 (A) Saxicolous thallus of *Rhizocarpon riparium* showing black bordering prothallus (× 12). (B) Saxicolous thallus of *Acarospora hilaris* showing development of areoles and lack of marginal lobation (× 12). (C) White non-lichenized prothalloid margin of *Catillaria sulphurata*; lichenized thallus is dark mottled (× 12). (D) Saxicolous thallus of *Rinodina oreina* showing marginal lobation and adjacent black spore mats (× 12).

Fig. 1.6 Umbilicate thalli of *Lasallia papulosa*.

Fig. 1.7 Foliose thallus of *Parmelia stuppea*.

Plate 5 Fruticose thalli of *Ramalina farinacea*.

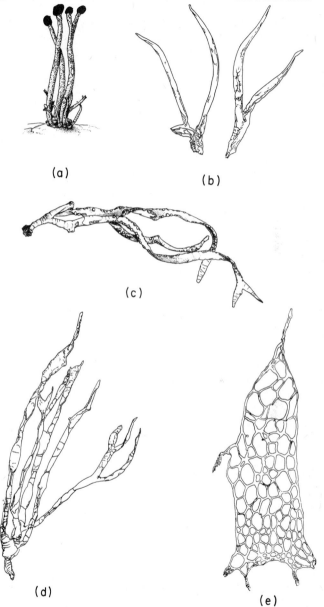

Fig. 1.8 Examples of fruticose lichens : (a) *Pilophorus acicularis* ; (b) *Thamnolia vermicularis* ; (c) *Roccella fuciformis* ; (d) *Ramalina homalea* ; (e) *R. reticulata* (near natural size). (Drawing by N. Halliday)

directly exposed; rhizines are lacking or only poorly developed. This growth form is characteristic of species of *Psora* and *Dermatocarpon* inhabiting soil, and the primary thallus of a number of *Cladonias* (Fig. 1.10). Individual squamules are relatively small, compared with foliose forms, usually no more than 5–10 mm in length and, lacking further anatomical refinements such as a lower cortex and rhizines, appear to be arrested in development.

Foliose lichens

The typical foliose thallus is a leafy dorsiventral plant body with a distinct lower cortex and rhizinal attachment to the substrate (Fig. 1.7; Plates 1, 8, 16). This highly structured growth form allows a far greater range of size and branching of lobes than is possible in crustose or squamulose groups.

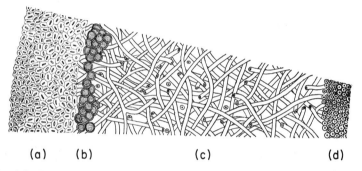

(a)   (b)   (c)   (d)

Fig. 1.9 Radial cross-section of a portion of the thallus of *Usnea densirostra*: (a) cortex; (b) algal layer; (c) medulla; (d) central cord (× 360). (From Grassi[104])

Some of the largest foliose lichens may be 0.3 m or more in diameter. Umbilicate lichens, the rock tripes, show a modification of the foliose growth form in which the thallus is fully corticated but circular in outline and firmly attached to the substrate by a central umbilicus (Fig. 1.6). This unusual form has developed independently in several totally unrelated groups, both among the Pyrenomycetes (*Dermatocarpon*) and the Discomycetes (peltate *Lecanoras*, *Omphalodium* and the Umbilicariaceae).

Fruticose lichens

Fruticose lichens are hair-like (Plate 5), shrubby, finger-like, or strap-shaped (Figs. 1.8, 1.10) with a wide range in size from minute *Cladonias* only 1 or 2 mm high to strands of *Usnea* up to 5 m long.[127] The internal structure is radial with a dense outer cortex, a thin algal layer, a medulla and a more or less hollow centre or a dense central cord (Fig. 1.9). The thallus may be round or flattened, unbranched or richly branched. Weak

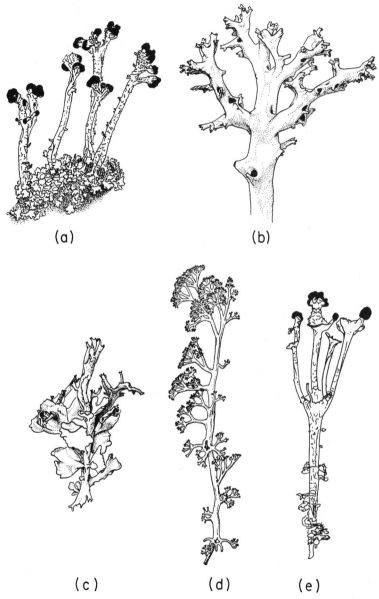

(a)   (b)

(c)   (d)   (e)

Fig. 1.10 Types of podetia in the genus *Cladonia* : (a) *Cladonia cristatella* ;
(b) *C. perforata* ; (c) *C. turgida* ; (d) *C. rangiferina* ; (e) *C. gracilis* (natural size).
(Drawings by N. Halliday, except d from Asahina[25])

dorsiventral differentiation is evident in *Evernia* which is transitional to the foliose genus *Pseudevernia* (Plate 16D). Fruticose lichens are anchored by basal rhizoidal strands derived from the cortex[217] although many species are without any attachment to the substrate. The fruticose structures of the common soil lichen *Cladonia* are called podetia (Fig. 1.10). These are hollow and differ from other fruticose lichens in being derived ontogenetically from tissues peripheral to the ascogonial complex in the primary squamules. By contrast, the solid fruticose structures of *Stereocaulon*, called pseudopodetia, have a thalline origin.[166]

VEGETATIVE STRUCTURES

Lichens are characterized by a variety of vegetative structures. Some, including rhizines, tomentum and cilia, are also known among fungi.

Fig. 1.11 Formation of soredia in *Nephroma parile* (magnified). (From Moreau[200])

Soredia, isidia, hormocysts, lobules, cyphellae and pseudocyphellae, and cephalodia, however, are not produced by non-lichenized fungi, algae, or for that matter by the isolated lichen fungus. They are unique products of the composite thallus. Many serve as vegetative diaspores, organs of dispersal, while some are presumed to have physiological functions. All are of great value in differentiating species of lichens.

Soredia

Soredia are separable non-corticated clumps of a few algal cells closely enveloped by hyphae. Individual soredia are of the order of 25–100 μm in diameter (Plate 6B) but gain in size when several adhere together in sub-macroscopic granular masses (Plate 6A). They originate in the medulla and algal layer, apparently following an overgrowth of the algae, and erupt through pores or cracks in the cortex (Fig. 1.11). The entire thallus of

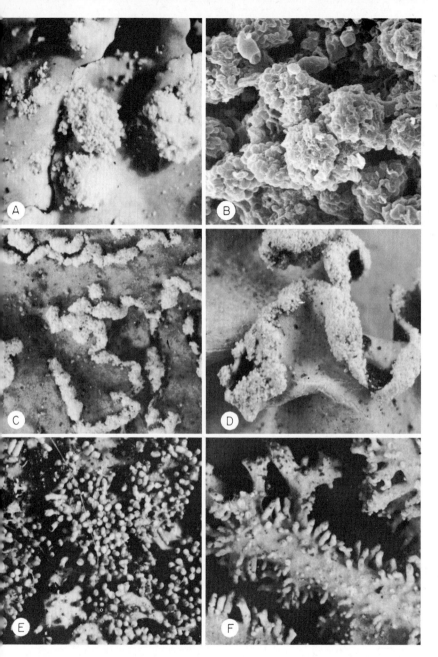

Plate 6 (A) Laminal soralia of *Parmelia aurulenta* (× 15). (B) Soredia of *Physcia pusilloides* (× 450). (C) Marginal soralia of *Anaptychia domingensis* (× 12). (D) Marginal soralia of *P. cristifera* (× 12). (E) Isidia of *P. ecaperata* (× 15). (F) Isidia of *Pseudevernia furfuracea* (× 12).

primitive crustose lichens such as *Lepraria* is essentially a continuous layer of diffuse soredia (Plate 1A).

Delimited masses of erupted soredia, called soralia, are oriented on the thallus in shapes characteristic for each lichen species. They are broadly classified as marginal (Plate 6C, D) or laminal (Plate 6A), depending on their position on the thallus, and farinose or granular according to their fineness. Soralia are especially common in foliose and fruticose lichens but comparatively rare in crustose groups (Table 3.1, p. 40). Hormocysts, known so far only in the gelatinous lichen *Lempholemma*,[77] are soredia-like lichenized algal structures that originate in apothecia-like hormocystangia. They apparently have the same function as soredia.

There are a number of non-sorediate lichens that are indistinguishable from sorediate species, except for the lack of soredia. Some well-known examples are listed in Table 1.1. This parallelism suggests that sorediate forms may have arisen from non-sorediate parent species more or less at random. Taxonomists, however, are in agreement that soredia behave as genetically determined, species-constant traits.[216A] What are the causes of soredial production? Shading, increased moisture and other environmental factors have been suspected since they would stimulate algal growth and, as an after effect, soredial formation,[156] but there is no experimental proof of this.

Table 1.1 Pairs of lichen species which differ only in the presence or absence of soredia.

Non-sorediate	Sorediate
Caloplaca elegans	*C. sorediata*
Evernia esorediosa	*E. mesomorpha*
Letharia columbiana	*L. vulpina*
Parmelia cetrata	*P. reticulata*
Parmelia latissima	*P. cristifera*
Parmelia zollingeri	*P. dilatata*
Physcia ciliata	*P. orbicularis*

Isidia

Isidia are cylindrical, finger-like protuberances of the upper cortex in which algal and fungal tissues are more or less continuously incorporated (Plate 6E, F). They are an integral part of the thallus, though often fragile and easily broken off, and are produced uniformly over the upper surface without any unique patterns of orientation. They range from 0.01 to

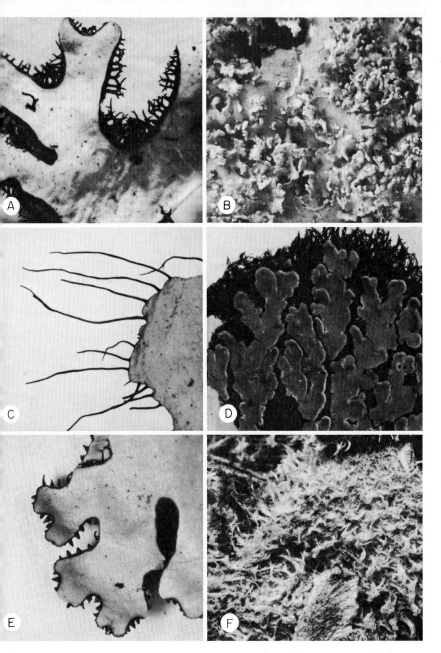

Plate 7 (A) Dichotomously branched rhizines of *Parmelia exsecta* (× 12). (B) Marginal lobules of *Physcia lacinulata* (× 12). (C) Marginal cilia of *Parmelia rampoddensis* (× 6). (D) Thallus and bordering hypothallus of *Pannaria mariana* (× 6). (E) Marginal bulbate cilia of *Parmelia abstrusa* (× 12). (F) Tomentum on the lower surface of *Erioderma chilense* (× 12).

0.3 mm in diameter and from 0.5 to 3.0 mm in height. While acting primarily as vegetative propagules, isidia also increase the surface area and presumably the assimilative capacity of the thallus, but it is not apparent that isidiate species hold any advantage over non-isidiate species. Up to 25–30% of the species in foliose and fruticose genera have isidia (Table 3.1, p. 40) but they are much rarer in crustose groups. The causes of formation of isidia are unknown but, as with soredia, appear to be genetically determined.

Lobules

Lobules are, in general, any adventive growths from the lichen thallus, often originating along the margins of lobes. They are especially common and well developed in the foliose genera *Anaptychia*, *Nephroma*, *Parmelia* and *Peltigera* (Plate 7B). They intergrade to a certain extent and may be-

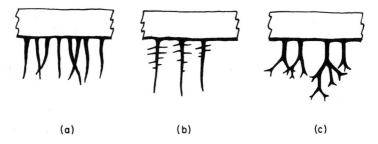

| (a) | (b) | (c) |

Fig. 1.12   Types of branching in rhizines of *Parmelia* : (a) simple ; (b) squarrose ; (c) dichotomous (enlarged). (From Hale[124])

confused with isidia but differ in being dorsiventral. The lobules of *Peltigera* are of special interest since they have been shown experimentally to be regenerative and stimulated by tearing or wounding the cortex.[263] In all groups lobules appear to be effective vegetative propagules.

Rhizines

Rhizines are compacted strands of colourless or blackened hyphae that originate largely from the lower cortex and anchor the thallus to the substrate.[217] The simplest rhizines are unbranched, as in *Cetraria*, *Physcia* and many *Parmelias* (Fig. 1.12a). Branching rhizines are of two kinds, squarrose, as seen in *Anaptychia* and some species of *Parmelia*, and dichotomous, in other *Parmelia* species (Fig. 1.12b, c; Plate 7A). To what extent rhizines are capable of transporting dissolved mineral or organic metabolites from the substrate to the thallus has not yet been established.

Tomentum

Tomentum differs from rhizines in lacking compaction. It is a felty, hirsute or cottony mat of multicellular rhizoidal chains or loosely organized strands (Plate 7F; Plate 8A), commonly occurring in forms lacking or with only a poorly developed lower cortex, as the Collemataceae, Peltigeraceae and Stictaceae. Tomentum is also characteristic of the upper surface in many species of *Peltigera* and *Erioderma* and other genera. The blackened hypothallus of *Coccocarpia*, *Pannaria* (Plate 7D) and a few other genera, is probably closely related to tomentum.

Cilia

Cilia are hair-like thalline appendages, decolourized or carbonized strands of hyphae that originate along the lobe margins (Plate 7C) or on the exciples of lecanorine apothecia in *Usnea*. A peculiar modification with a bulbate inflated base is known in *Parmelia* (Plate 7E). Cilia appear

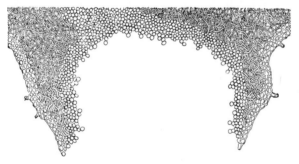

Fig. 1.13 Vertical cross-section of a cyphella in *Sticta* (magnified). (From Grassi[104])

to be related to rhizines, although in view of their diversity they may well have originated in several ways. As a rule, cilia are restricted to the more or less structurally advanced foliose genera such as *Anaptychia*, *Cetraria* and *Parmelia*, and the fruticose families. They are virtually unknown in crustose and gelatinous lichens, and the large foliose families Peltigeraceae and Stictaceae.

Cyphellae and pseudocyphellae

Cyphellae are rather large round pits in the lower cortex found only in the foliose genus *Sticta*; they have a cellular structure (Fig. 1.13; Plate 8A, B). Pseudocyphellae are smaller, rarely exceeding 1 mm, and less differentiated, often no more than an irregular opening in the cortex through which medullary hyphae protrude (Plate 8E, F). These pores occur on either the upper or lower cortex and are conspicuously developed in the genera *Cetraria, Cetrelia, Parmelia* and *Pseudocyphellaria*. It is believed that these two types of pores function in gas exchange and aeration of the thallus, although this has never been specifically demonstrated.

The size of pseudocyphellae is quite constant in most species and has considerable taxonomic importance. This fact was first recognized in the arctic-alpine fruticose lichen *Cornicularia divergens* and has recently helped to separate populations of *Cetrelia olivetorum* and *Cetrelia chicitae*, two extremely similar foliose lichens in eastern North America.[73B] At first thought to be indistinguishable when sterile except by chemical tests (olivetoric and alectoronic acids, respectively),[68] these two species differ significantly in pore size, the average maximum size in *C. olivetorum* being 0.48 mm and in *C. chicitae* 1.20 mm (Plate 8E, F).[127]

Cephalodia

Cephalodia are outgrowths of the thallus containing a 'foreign' alga different from the host phycobiont. In *Peltigera aphthosa* they appear to begin when a *Nostoc* colony falls on the thallus surface and is enmeshed by aerial hyphae (Plate 8C). The small discrete thallus that develops is about a millimetre or more in diameter (Plate 8D). Since most cephalodia contain the blue-green *Nostoc*, they may be able to fix nitrogen and thus provide the host thallus with this important element. They do not seem to have any detrimental effects on the host.

Cephalodia are borne superficially on the thallus of about 100 species of *Peltigera, Lecanora, Stereocaulon, Lecidea* and several smaller crustose genera. By no means do all the species in these genera produce cephalodia, but they are constant features for those species in which they occur. Internal cephalodia, colonies of blue-green algae distinct from the main algal layer, are found in some species of *Nephroma, Lobaria* and *Solorina*. The 'isidia' of *Peltigera evansiana* and *P. lepidophora* are in reality tiny isidioid cephalodia since they are easily dislodged from the host thallus.[179]

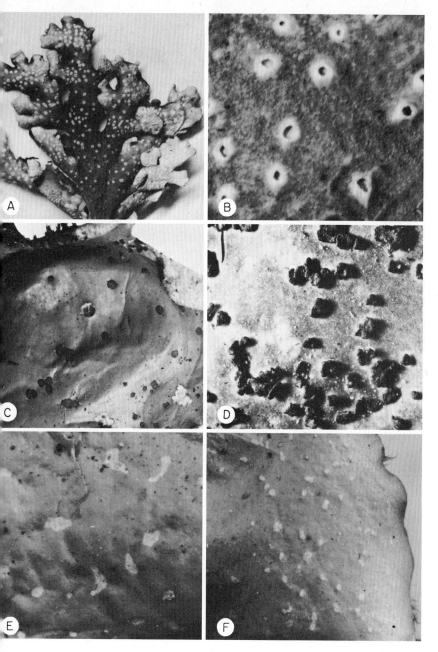

Plate 8 (A) Lower surface of *Sticta gracilis* (× 2). (B) Close-up of cyphellae
of *S. gracilis* (× 15). (C) Upper surface of *Peltigera aphthosa* (× 2). (D) Close-up
of cephalodia of *P. aphthosa* (× 15). (E) Pseudocyphellae in the upper cortex of
Cetrelia chicitae (× 15). (F) Pseudocyphellae in the upper cortex of *Cetrelia
cetrarioides* (× 15).

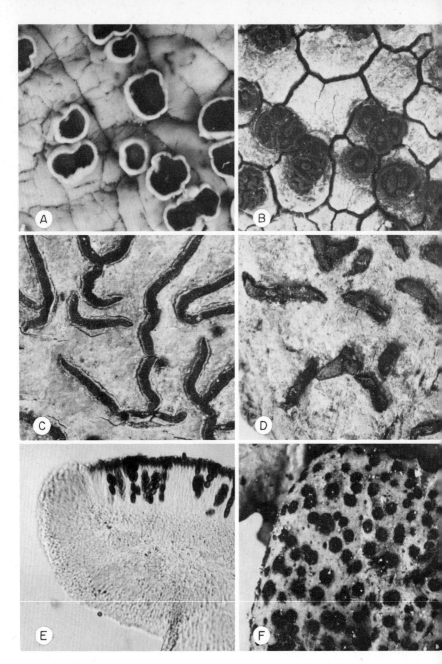

Plate 9 (A) Lecanorine apothecia of *Lecanora arizonica* (× 12). (B) Lecideine apothecia of *Lecidea ruboniza* (× 12). (C) Lirelliform apothecia of *Phaeographina dendritica* (× 12). (D) Lirelliform pseudothecia of *Opegrapha varia* (× 12). (E) Hymenium of *Physcia aipolia* (× 150). (F) The parasymbiont *Abrothallus parmeliarum* on *Platismatia glauca* (× 12).

2

Morphology of Reproductive Structures

The life cycle of fungi is completed when the vegetative thallus produces fruiting bodies that contain spores. The spores are disseminated and under suitable conditions germinate to form new thalli. Fungi are divided into two major groups according to the method of spore production: one, the Ascomycetes, where typically eight spores develop within a sac, the ascus; the other, the Basidiomycetes, where four spores are produced externally on a club-shaped basidium. The overwhelming majority of lichens are Ascomycetes with fruiting bodies (ascocarps) essentially identical with those of non-lichenized fungi, except that those of lichens tend to be perennial and more durable, sporulating over a period of several years. The lichenized Basidiomycetes make up a very small group of lichens.

ASCOMYCETES

The Ascomycetes were formerly considered to be a relatively homogeneous group characterized by ascocarps containing a fruiting layer (the hymenium) of spore-containing asci and sterile thread-like branched or unbranched paraphyses (Fig. 2.2; Plate 9E). In 1932, however, Nannfeldt,[204] a Swedish mycologist, summarized our knowledge of the ontogeny and structure of the hymenium and concluded that there are (excluding the Plectascales, *Aspergillus* and allied genera) two fundamentally different types, designated as ascohymenial and ascolocular. In the ascohymenial groups, the ascogenous system or ascogone is initiated on the mycelium unprotected by other hyphae. Supporting cells of the ascogone give rise to true unbranched paraphyses with free tips, and hyphae adjacent to the ascogonial complex develop into the excipular tissues (rim) of the mature ascocarp. In the ascolocular type of development, the first formed structure is a specialized vegetative stroma, a layer of sterile hyphal tissue within which the ascogenous system is initiated. The locules in the stroma are separated by branched pseudoparaphyses.[106]

Although his work was directed primarily at non-lichenized fungi, Nannfeldt included enough lichens to prove that they too have similar hymenial structures. For example, *Opegrapha*, a genus long thought to be related to *Graphis*, has lirelliform 'ascocarps' (Plate 9D) that are ascolocular. The lirelliform apothecia of *Graphis* and other genera in the Graphidaceae (Plate 9C) are ascohymenial.

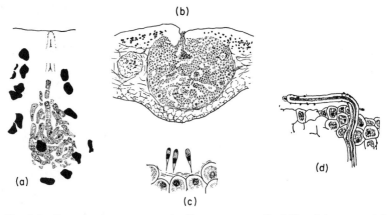

Fig. 2.1 Reproductive structures in *Dermatocarpon fluviatile* : (a) ascogonial coil ; (b) cross-section of a pycnidium ; (c) microconidia budded from the pycnidial wall ; (d) trichogyne emerging at the upper cortex with adhering microconidia (magnified). (From Stevens[256])

The American mycologist Luttrell[188] showed by a more detailed study that bitunicate asci, those with a double wall structure, are correlated with ascolocular development and unitunicate, with a single wall structure, with the ascohymenial. Chadefaud[61] provided additional correlation with the apical structure of the ascus, in general nassasceous (with rod-like amyloid bodies) in bitunicate species, annelasceous (with ringed amyloid bodies) in unitunicate species.

Although extremely few fungi have been given adequate study, almost all mycologists accept the validity of the distinction between ascohymenial and ascolocular fungi. Groenhart,[106] however, found discrepancies in these correlations and concluded that the only infallible criterion is in the initiation of the ascogenous system, a character that must be determined early in culture. There are examples of species which have ascohymenial ascocarps but which appear to have an ascolocular origin. It has not yet been decided whether the abnormally thickened indistinctly two-walled ascus of many lichens is bitunicate or unitunicate. It is unfortunate that neither the appearance of the hymenium at maturity nor ascal structure necessarily reflect or correlate with the initial ontogenetic development of the ascocarp. Richardson[222A] has ably summarized recent developments here.

Ascohymenial lichens

The ontogeny of the ascocarps is known for only a few species of lichens and varies from family to family. The overall processes can be briefly outlined as follows with allowance for considerable oversimplification. The

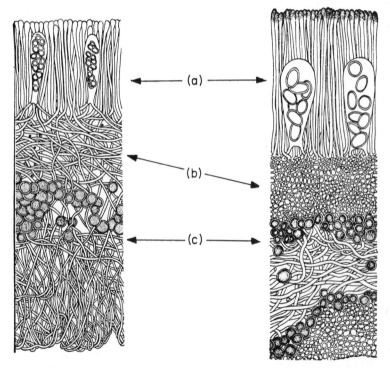

Fig. 2.2 Sections of the apothecial discs of *Sticta kunthii* (left) and *Parmelia microsticta*: (a) hymenium with paraphyses and asci; (b) hypothecium; (c) medullary and algal layers (magnified). (From Grassi[104])

first sign of ascocarp development is an ascogonial coil or simply a compacted group of cells in the medulla or lower algal layer (Fig. 2.1a), usually

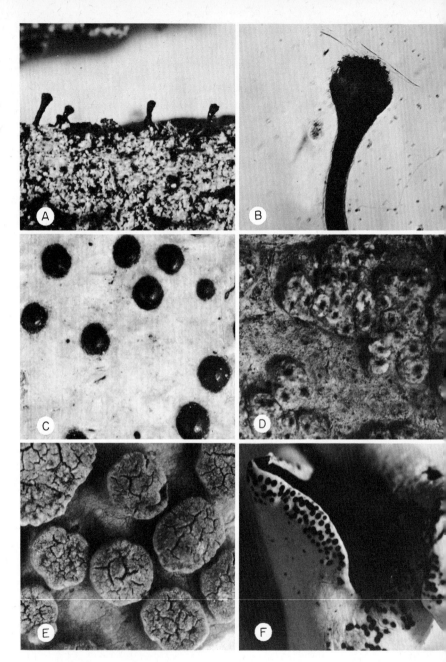

Plate 10 (A) Thallus and mazaedia of *Chaenotheca chrysocephala* (× 15).
(B) Mazaedium of *C. chrysocephala* (× 150). (C) Perithecia of *Pyrenula nitida*
(× 12). (D) Aggregated pseudothecia of *Melanotheca aggregata* (× 12).
(E) Apothecia-like pseudothecia of *Roccella fimbriata* (× 10). (F) Laminal
pycnidia of *Parmelia explanata* (× 15).

identified with histological stains, to which a long trichogyne is connected, extending to the thallus surface and protruding 10–20 μm (Fig. 2.1d). Ascogonia often occur in clumps and are uninucleate or multinucleate.[87] After the trichogyne degenerates, the ascogonial tissues proliferate into a mass of binucleate cells from which ultimately the asci are derived. Tissues peripheral to the ascogonial complex contribute to the formation of the hypothecium, paraphyses and excipular structures (Fig. 2.5). The mature ascocarps are classified as apothecia or perithecia.

Apothecia

Apothecia typically are open disc- or cup-shaped structures (Plate 9). The hymenium is a thin layer of asci and paraphyses lining the inner surface of the cup (Fig. 2.2). Spores are ejected from the asci forcibly by

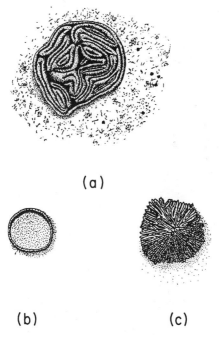

(a)

(b)   (c)

Fig. 2.3 Variation in the apothecial disc of *Umbilicaria* : (a) regularly gyrose ; (b) smooth lecideine ; (c) radially gyrose (enlarged). (From Kershaw[158])

pressure of swollen moistened cell walls. There is a small group of lichens classified as the order Caliciales in which the asci disintegrate at maturity and spores are liberated free in the hymenium (Plate 10A, B). This so-called mazaedium is similar to the ascocarp of the fungal family Onygenaceae.

The variety of shapes in apothecia of lichens is not great. The discs are ordinarily open and small, as in comparable Helotiales,[80] somewhere in the range of 0·5–10 mm in diameter. Only in the Peltigeraceae and Parmeliaceae do the apothecia commonly exceed 10–20 mm. While normally laminal or marginal, apothecia are found on the lower surface of the thallus in *Nephroma*. The disc of the apothecia in the rock tripes (Umbilicariaceae) appears convoluted because of the intrusion of sterile tissue into the hymenial layer (Fig. 2.3).[141A] Stipes are developed in a few families. Those in *Pilophorus* (Fig. 1.8a) and *Baeomyces* (Plate 2E) resemble the stipes of the non-lichenized genus *Leotia*.[166]

Most apothecia in lichens are gymnocarpous; that is, the hymenium develops fully exposed. Henssen[139] has recently traced consistent hemi-angiocarpous development in a group of genera that she combines in the family Peltigeraceae. In these lichens the hymenial layer completes most of its development below the cortex and only later expands and bursts the cortex so as to expose the disc.

Perithecia

Perithecia are immersed flask-shaped structures, usually not exceeding 1 mm in diameter. The hymenium lines the inner surface and spores are ejected or exuded through an apical pore. Perithecia are usually borne singly (Fig. 2.4c, d; Plate 10C) but become aggregated in stromatic clusters

(a)

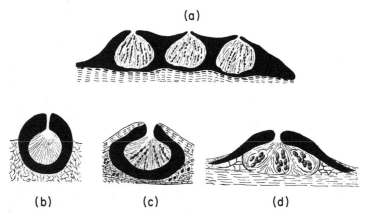

(b)   (c)   (d)

Fig. 2.4 Cross-sections of perithecia : (a) *Melanotheca diffusa* (stromatoid) ; (b) *Arthopyrenia saxicola* (pseudothecium) ; (c) *Pyrenula nitida* ; (d) *P. coryli* (magnified). (From des Abbayes[1])

in *Laurera*, *Melanotheca* and related tropical genera, although there is evidence now that these structures are pseudothecia rather than perithecia (Fig. 2.4a; Plate 10D). Johnson[151] has shown conclusively that the so-called stroma of the lichenized Pyrenomycetes is nothing more than elevated bark cells that have been modified by the lichen hyphae. The definition of the term stroma, however, has not yet been settled even among mycologists, and more definitive studies on this structure in lichens are needed.

Modifications of ascocarps

The presence of both thalline (hyphae and algae together) and purely fungal tissue introduces modifications in ascocarps that are unique to lichenized fungi and differentiate them from other fungi. Many lichens, of course, follow the same course of development as non-lichenized fungi, in which the apothecia and perithecia do not incorporate thalline tissue. In

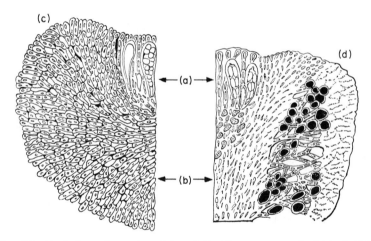

Fig. 2.5 Vertical sections of exciples of *Buellia parasaema* (lecideine) (left) and *Lecanora subfusca* (lecanorine) : (a) hymenial layer ; (b) hypothecium ; (c) fastigiate proper exciple ; (d) thalline exciple (magnified). (From Dughi[81])

some apothecia, the disc is surrounded by a rim, the proper exciple, which has a fastigiate cortex and is usually similar in colour to the disc (Fig. 2.5; Plate 9B). This condition, called lecideine, is characteristic of many common lichen families, the Lecideaceae, Cladoniaceae and Umbilicariaceae, to cite a few examples. Several groups with a proper exciple, including *Parmeliella*, *Placynthium* and *Polychidium*, have a prosoplectenchymatous cortex and are called superlecideine.

When thalline tissue enters into the formation of the apothecium, it appears as a second rim around the disc, of the same colour as the thallus (Plate 9A). Fully developed, this thalline exciple contains algae and all but displaces the proper exciple (Fig. 2.5). This so-called lecanorine apothecium characterizes many large lichen families, the Lecanoraceae, Parmeliaceae, Teloschistaceae and Usneaceae, as well as smaller groups. In a few species the thalline as well as proper exciples are nearly equal in size and both persist. Dughi[81] has proposed a detailed classification of lecanorine apothecia that can be consulted by those interested.

The elongate lirellae of the family Graphidaceae are now regarded as a kind of modified apothecium (Plate 9C). The carbonized walls are derived from thalline tissue. The primary perithecial wall of many Pyreno-mycetes is surrounded by a carbonized thalline layer, called the involu-crellum, which forms a double wall system useful in the taxonomy of these groups.[260]

Ascolocular lichens

The ascolocular lichens differ from ascohymenial lichens, as indicated above, in lacking a true hymenium and in having two-walled asci scattered in locules of the stroma. The thin outer ascal wall ruptures towards maturity, the inner wall extending to form a cylindrical sac.[223] The spores are thus carried above the general level of immature asci and released one at a time through a rather indistinct elastic apical pore (Fig. 2.6). The most

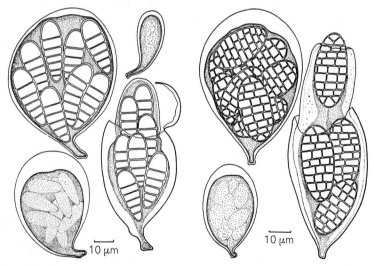

Fig. 2.6 Bitunicate asci of *Arthonia illicina* (left) and *Arthothelium spectabile* (magnified). (From Richardson and Morgan-Jones[223])

primitive ascolocular lichens, the tropical cryptothecioid genera,[106] have no organized fruiting bodies but are closely related to the non-lichenized Myrangiales. Asci are produced free on the thallus surface. Organized, though often weakly delimited, fruiting bodies are referred to as pseudo-thecia or ascostromata. They usually superficially resemble perithecia (Plate 2D) but also mimic apothecia in *Roccella* (Plate 10E) and lirelliform apothecia in *Opegrapha* (Plate 9D). Because of their extremely close

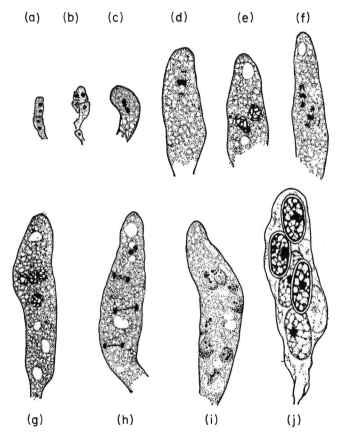

(a)   (b)   (c)   (d)   (e)   (f)

(g)   (h)   (i)   (j)

Fig. 2.7 Stages in the development of asci in *Dermatocarpon fluviatile*: (a) dicaryotic stage of the ascogenous hyphae; (b) crozier formation; (c) fused nuclei (zygote); (d) first meiotic division; (e) two-nucleate stage; (f) second meiotic division; (g) four-nucleate stage; (h)-(j) formation of spores by mitotic divisions (magnified). (From Stevens[256])

external similarity, pseudothecia and true ascocarps are easily confused even by experienced lichenologists. Ordinarily the demonstration of branched pseudoparaphyses and bitunicate asci is sufficient evidence for ascolocular lichens, but much work remains to be done in this difficult area of research.[222A]

Spores

Spores are the result of meiotic and equational division of the zygote in the primary ascus. Fusion of nuclei in early ascal formation has not been observed directly in lichen ascocarps but suggestive meiotic figures make it seem plausible (Fig. 2.7). Chromosome numbers determined during these stages range from $n=2$ in *Lecidea crustulata*[17] to $n=6$ or 8 in *Dermatocarpon fluviatile*.[256] The number of spores is most commonly 8,

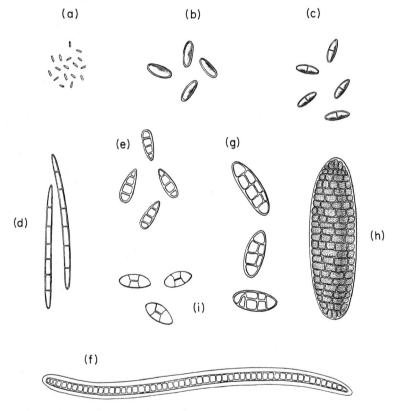

Fig. 2.8 Types of spores in lichens : (a), (b) simple ; (c) - (f) transversely septate ; (g), (h) muriform ; (i) polarilocular (magnified). (From Hale[120])

although *Acarospora, Candelaria, Sarcogyne* and *Anzia* typically have 64 or more. *Mycoblastus, Lopadium* and *Pertusaria*, among others, often have only one or two. The smallest spores are probably those of *Acarospora*, barely exceeding 1 μm. The largest recorded spores are 510 μm long in the tropical foliicolous lichen *Bacidia marginalis*.[232] As in the closely related non-lichenized order Helotiales, the spores are unornamented, although exospore sculpturing is reported in *Tholurna* and in the Caliciales.[264A] Spores are either colourless or brown.

Septation is one of the most useful features in separating spores. They are classified as follows (Fig. 2.8):

1 Simple spores: unicellular and unseptate, often small and thin-walled (*Lecidea, Lecanora, Parmelia, Usnea*), more rarely very large (to 300 μm) and thick-walled (*Pertusaria*).
2 Transversely septate spores: elongate and multicellular with 1 to as many as 30–40 transverse cross walls (*Catillaria, Graphis, Pyrenula*).
3 Muriform spores: multicellular with both transverse and longitudinal walls, often large (*Phaeographina, Lopadium, Umbilicaria, Diploschistes*).
4 Polarilocular spores: two-celled spores with a thick median wall and a thin isthmus, or conversely a single-celled spore with a median constriction (Teloschistaceae).

Pycnidia

Pycnidia (spermagonia) are extremely common in lichens and seem more closely related to those of rust fungi than to those of the Ascomycetes. They are flask-shaped (Fig. 2.1b), immersed or superficial structures that resemble perithecia (Plate 10F). If we take *Dermatocarpon fluviatile* as an example,[256] the first indication of a pycnidium is an aggregation of darker staining cells in the lower portion of the algal layer. This cluster increases in size until the upper cortex bursts and a pore is formed. Uninucleate bacilliform microconidia (spermatia) are budded off more or less continuously from specialized cells (conidiophores) lining the pycnidial cavity and exuded through the apical pore (Fig. 2.1c). There is no seasonal development but senescence is finally reached, discharge of microconidia ceases, and the pycnidium remains as an empty chamber.

In the gelatinous·family Lichinaceae the ascocarps develop within the pycnidia.[139] The ascogones give rise to conidiophores as well as to ascogenous hyphae and asci and all are intermingled at maturity in the hymenium. These so-called pycnoascocarps are considered to be primitive. Microconidia in the gelatinous lichen *Collema* may be produced internally free in the medulla rather than in distinct pycnidia.[78]

Microconidia are unicellular and tend to be extremely small, only 1–5 μm long. Very large septate conidia have been called stylospores or macroconidia but should not be confused with the conidia of the Imperfect Fungi. In lichens they apparently have the same origin as microconidia.[232] Nylander placed great emphasis on the use of microconidia in lichen taxonomy. For example, he segregated *Parmeliopsis*, a small group of foliose species with exobasidial conidiophores (microconidia produced terminally), from *Parmelia*, which has endobasidial conidiophores (microconidia produced laterally and terminally). Length and bending of microconidia are also important characters in the Physciaceae. Recent lichenologists, however, have given much less attention to microconidia.

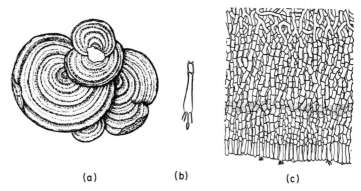

(a)   (b)   (c)

Fig. 2.9 Structure of the lichenized Basidiomycete *Cora pavonia* : (a) thallus (× 1) ; (b) basidium and basidiospores ; (c) section of the hymenium on the lower surface (magnified). (a, from des Abbayes[1] ; b, c from Grassi[104])

BASIDIOMYCETES

In contrast to the thousands of species of lichenized Ascomycetes there are less than twenty species of lichenized Basidiomycetes.[267] Oberwinkler[208A] has classified these broadly into three types, corticioid, clavarioid, and agaricoid, corresponding to three well-known fungal groups. *Cora*, a representative corticioid type, has a bracket-shaped thallus (Fig. 2.9) with a distinct internal algal layer; basidiophores and basidia are produced on the lower surface. The last two types include several species, some common in Europe, which attack *Coccomyxa* algae and establish a successful symbiotic union (as in squamulate *Omphalina ericetorum*) or behave as parasites (as in *O. grisella*[216B]). Some lichenologists consider these to be lichenized fungi rather than true lichens. The reproduction and development of all these Basidiomycete forms are presumably very similar to those of their non-lichenized relatives.

3

Reproduction and Dispersal

The reproduction of lichens is a complex, little-understood process that involves both sexual and asexual stages. Because of the slow growth of the thalli and the lack of success in inducing fructification in mycobiont cultures, no other aspect of lichenology is as poorly known and intractable as reproduction. No lichen has been satisfactorily studied in the field or in the laboratory. Studies of observable structures in the thallus and comparison with those of better known non-lichenized fungi give hints as to the probable course, but most of our conclusions are still speculative.

VEGETATIVE DIASPORES

Vegetative diaspores are lichenized structures in which the algal and fungal components together act as separable autonomous sub-units of the thallus, similar in function to gemmae in *Marchantia* or the plantlets of the ornamental *Kalanchoë*. On breaking free from the thallus, they are capable of continuing existence. In lichens these diaspores are soredia (Plate 6 A–D), isidia (Plate 6E, F), the rare hormocysts, various thalline structures such as squamules, lobules (Plate 7B), fragments, or even the whole thallus; they may be transmitted by water, wind, animals or birds.[36A] There is no general summary on the number of lichens that have these diaspores, although we can arrive at an accurate estimate of the frequency of soredia and isidia in some of the major genera by examining taxonomic monographs (Table 3.1). The frequency varies greatly from genus to genus, but foliose and fruticose groups incline to have numerous and well developed diaspores, crustose groups relatively few.

Table 3.1   Percentage of isidiate and sorediate species in selected lichen genera. *Anaptychia*, *Collema*, *Parmelia* and *Physcia* are foliose ; the remaining groups are crustose or subcrustose.

	Number of species	% Isidiate	% Sorediate
Anaptychia (world)[162]	79	6	16
Collema (Europe)[78]	35	46	0
Parmelia subg. *Amphigymnia* (world)[124]	106	23	32
Physcia (North America)[264]	47	9	49
Foliicolous lichens (world)[232]	236	2	1
Lecanora (in part) (world)[190]	118	0	1
Lecidea (United States)[187]	88	0	3
(Europe)[192]	235	0	9
Lichinaceae (world)[139]	32	0	0
Maronea (world)[190]	13	0	0

The wide geographic distribution of many lichens attests to the efficiency of their means of reproduction, whatever it may be. A comparative lack of endemism is now thought to be characteristic of lichens. For example, 26 of the 106 species of the broad-lobed *Parmelias* are widely distributed on all continents.[124] For higher plants quite the contrary would be true; most species show a high degree of restriction to small geographic areas. To continue with *Parmelia*, 21 (81%) of the 26 pantropical or pantemperate species have soredia or isidia.[124] This fact alone would not be significant were it not that less than 50% of the species endemic to a single continent produce vegetative diaspores. As a more specific example, *P. cristifera*, one of the commonest pantropical lichens, is sorediate (Plate 6D). *P. latissima*, which lacks soredia, but is indistinguishable from *P. cristifera* in all other respects, has been reported only from the tropical New World with the exception of one collection in India (Fig. 3.1).

One usually finds, in addition, that sorediate and isidiate species have greater numerical abundance in lichen communities than their simple frequency in a genus would dictate. Of the 18 statistically most frequent lichens on deciduous trees in southern Wisconsin, for example, 13 are sorediate or isidiate and only 5 lack these diaspores. Of the 18 rarest species in the same area, 16 lack any obvious vegetative diaspores.[111] Degelius[79] calculated that 84% of the 56 species that invade twigs of *Fraxinus excelsior* have diaspores.

How lichens become dispersed so effectively is one of the major unsolved problems in lichenology. It is no easy task to identify, tag and record the growth of submacroscopic vegetative propagules in nature. Tobler[265] sowed soredia of *Cladonia* species on sterile soil in the laboratory and was able to observe early stages of development for several months. Nienburg[206] pieced together a sequence of developmental stages in *Pseudevernia furfuracea* from individual isidia to mature thalli on datable twig internodes, but other than this composite type of study, there is surprisingly little direct evidence on how fast vegetative propagules grow, whether they regularly develop into mature lichen thalli, and what effect this mode of reproduction has on genotypic variation. In spite of these uncertainties, it is an inescapable conclusion that lichens rely on vegetative reproduction to a very large, though undetermined, extent.

Fig. 3.1 World distribution of sorediate *Parmelia cristifera* (area between solid lines) and non-sorediate *P. latissima* (area within pecked line). (Adapted from Hale[124])

SEXUAL REPRODUCTION

Sexual reproduction in micro-organisms is usually associated with spore production. We should not say, therefore, that the composite thallus reproduces sexually, because spore production is a characteristic of the fungus alone, though it may be functioning in a symbiotic state with algae. A lichen thallus is reconstituted secondarily through lichenization, by the combining of germinated ascospores and a suitable alga. Microconidia, though not produced by sexual processes, represent another source of non-lichenized diaspores, which in certain fungi may germinate and form new vegetative thalli. Among lichens, however, there has been little success in germinating microconidia except for the work by Möller[198] in 1887 who cultivated identical mycelia from the ascospores and conidia of *Calicium parietinum*.

Sexual processes

It might be helpful to review the structures in lichens that could be expected to participate in sexual reproduction and their relation to those in non-lichenized fungi. These are the ascogonial apparatus, pycnidia and microconidia. Comparable organs are known in the closely related fungi of the order Helotiales where the contents of the ascogonial coil act as the female element and the microconidia function as a spermatizing element, although their exact role in sexual reproduction is not always evident. The female nucleus and the microconidia may originate on different mycelia, and plasmogamy occurs when a compatible microconidium pairs or fuses with the female nucleus. This heteromictic type of reproduction is by no means a common occurrence in non-lichenized fungi.[58] There seems to be a general evolutionary tendency toward a diminished role for microconidia and a simplification of plasmogamy. Homomictic plasmogamy, the fusion of genetically similar nuclei from the same thallus, is apparently a widespread phenomenon. The spermatizing element may be any hypha that can contact the ascogonium, or in extreme cases of reduction there is autogamous or parthenogenetic fusion of nuclei within the ascogonium itself. All of these types of plasmogamy have been observed in various degrees of complexity among non-lichenized Ascomycetes,[100] and lichenized fungi might be expected to show the same processes.

The steps leading to formation of ascocarps and spores in lichens, however, are very poorly known compared with other fungi. Ahmadjian[9] has cultured the mycobiont of *Cladonia cristatella* until both pycnidia and rudimentary podetia developed (Plate 11D). Ascogonia and trichogynes, but not asci, were demonstrated at the tips of the podetia. Other than this example, mycobionts of lichenized fungi have never been known to fructify under laboratory conditions. There is good reason, nevertheless, to suppose that lichens are homothallic and that microconidia are largely functionless, notwithstanding the fact that they have been seen adhering to the trichogynes in *Dermatocarpon fluviatile*,[256] *Stereocaulon nesaeum*[152] and other species. It is doubtful that anyone has actually seen a microconidium migrate to the ascogonial coil and fuse with the female nucleus. Many species of lichens lack pycnidia entirely and sexual reproduction, if it occurs at all, must of necessity proceed in their absence. According to Santesson,[232] two-thirds of the 234 species of crustose foliicolous lichens lack pycnidia yet regularly form apothecia. In addition, even when pycnidia are present, trichogynes may be lacking or degenerate and fail to reach the surface of the thallus.

Stages in the reproduction of the thallus

Let us assume for the moment that sexual reproduction could occur in

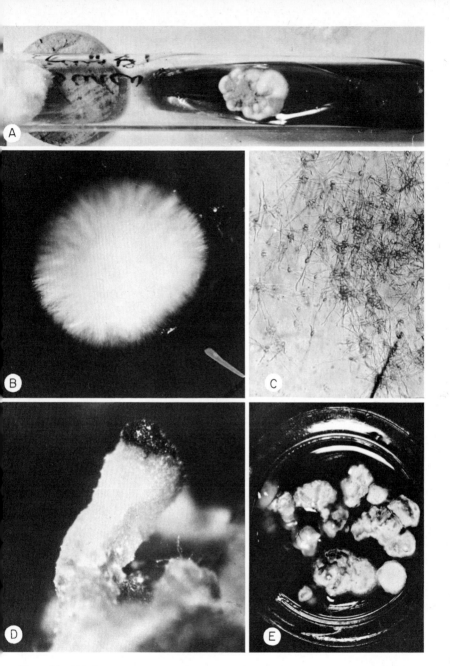

Plate 11 (A) Agar slant culture of the mycobiont of *Sarcogyne similis* (× 1). (B) Single colony of the mycobiont of *S. similis* showing aerial hyphae (about 3 mm in diameter). (C) Germinating spores of *Buellia stillingiana* (× 150). (D) Young podetium of the mycobiont of *Cladonia cristatella* (about 1 mm high). (E) Mycobiont of *B. stillingiana* in liquid synthetic medium. (D by V. Ahmadjian)

some lichens. What are the hypothetical stages that lead to the re-creation of a lichen thallus?

After an ascospore is ejected or exuded from the ascocarp, it falls back on the thallus surface or to the substrate. If moisture is adequate, the spores will germinate (Plate 11C), although the actual time needed for germination or duration of dormancy is impossible to state. Spores vary considerably when germinated on agar,[281] an unnatural substrate which may even inhibit germination. Crustose species may germinate in as little as one or two hours. Foliose species may take several days, while some species of *Peltigera*[240] and a few other lichens do not seem to have viable spores at all, at least under cultural conditions.

Once spores germinate, there are several ways in which further development can proceed. Germ tubes may rapidly grow into an extensive mycelial mat that covers the substrate (Plate 11A). This has apparently been observed in the foliicolous lichen *Strigula complanata*[277] and is suspected to be rather common among tropical corticolous lichens. How long the mycelium can maintain itself autonomously is unknown, but contact with algae is probably made rather quickly.

A second pathway is probably more typical of lichens in temperate and boreal areas where growth is slower. The spores, after being ejected, do not germinate at once but remain dormant for some time. The spores of *Xanthoria parietina*, for example, become attached to the substratum individually and are dormant until contact with algae is made.[282] The spores of saxicolous species often become clustered in rock crevices and adhere together as a distinct mat not much more than 0.1 mm wide. These mats appear to be characteristic of *Rinodina oreina* (Plate 4D) and other crustose prothalloid species. Each mat is equivalent to a primitive prothallus and has no contact with algae for months or perhaps even years.[41,156]

Should the spores germinate but fail to contact algae, they may simply perish. There is also the chance, however remote, that the germinated spores form a mycelium indistinguishable from that of non-lichenized fungi and enter a true state of saprophytism. They would have lost, at least temporarily, their dependence on algae. Evidence for this hypothetical reversion to a non-lichenized state obtains mostly from cultural studies of *Buellia stillingiana*. The spores of some strains of this corticolous crustose lichen readily germinate (Plate 11C) and form dense colonies that sporulate on agar.[117] The conidia are morphologically identical with those of a non-lichenized saprophyte, *Sporidesmium folliculatum* (Fig. 3.2), but it cannot be said at this time that *Buellia* and *Sporidesmium* are phases of the same fungus.

The similarity of the conidial genus *Pullularia* and the conidia of *Lecidea sylvicola* also suggests a dual existence for this lichenizable fungus.[6] Likewise, a free-living pycnidial stage, *Pycnidiella resinae*, is associated with the weakly lichenized *Biatorella resinae*, and these are perhaps stages of a single fungus. Nordin[207] reports that the parasymbiont *Abrothallus suecicus* is the perfect stage of the imperfect fungus *Phoma*. Other than these examples we are quite ignorant of the possible relation between conidial stages of non-lichenized fungi and the perfect stages of lichenized fungi.

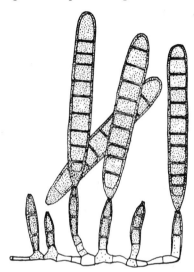

Fig. 3.2 Conidia of the mycobiont of *Buellia stillingiana* (magnified). (From Hale[117])

Lichenization

The lichen fungus, so far in this account, has succeeded only in establishing itself on a substrate; it is unrecognizable as a lichen. The next step would be to make contact with algae that are either already present in close proximity with the mycelium or are carried there by wind, water or animals. The hyphae do not seem to grow actively toward algae; the reaction is a simple thigmotropic response. If several kinds of algae are contacted, the fungus probably incorporates them indiscriminately into the prothallus. Because only one kind of alga, usually *Trebouxia*, is the phycobiont of most higher lichens, lichenization must be a highly selective process. Ahmadjian[4] is of the opinion that the common protococcoid algae are soon killed off by the fungus, since they are unable to establish and maintain a balanced symbiotic state. Usually only trebouxoid forms resist the fungus, perhaps in part because of their ability to become adapted to lower light intensity as well as their evident evolutionary dependence on symbiosis.

A fundamental unsolved problem is the source of symbiotic algae in nature. *Trebouxia* is extremely rare in algal covers on tree bark and rock and, in fact, no one has conclusively proved that *Trebouxia* is a free-living alga. Ahmadjian[10] theorizes that reports of this alga may actually be based on free *Trebouxia*-containing soredia or algal cells that have escaped from soredia. Germinating fungal spores might therefore depend on soredia or other vegetative diaspores as sources of *Trebouxia*. It is conceivable, too, that algal cells released from the thallus during heavy rain, either as free cells or motile spores, are available as phycobionts.

There are other unanswered questions about lichenization on which we can only speculate. How often do several ascospores from different lichens germinate together and become incorporated in a growing thallus? It is not uncommon for a small rock or bark surface to be inhabited by as many as 5 or 10 different lichens. If all of the ejected spores became enmeshed in a prothalloid mat, it would be logical to suppose that they have equal opportunity to participate in lichenization. The identity of this heterosporic mat could be further complicated by the incorporation of soredia with enclosed *Trebouxia* and other 'foreign' thallus fragments.[265] The resulting hybrid thallus should be a veritable Gordian knot of fused hyphae and vegetative anastomoses, yet a mature lichen thallus produces only one kind of ascocarp. It is even more intriguing that chemical features are unaffected by this conceivable fusion of different genetic strains. Thalli of substitution chemical strains (acid A or acid B) almost always contain only one or two acids, not a mixture of acid A and acid B. One fungal component in a polysporic thallus must have a highly efficient mechanism to suppress other components.

Development of the thallus

The subsequent steps in the formation of a fully differentiated thallus have been outlined by Werner[282] on the basis of careful studies of *Xanthoria parietina*, *Caloplaca vitellinula* and *Lecania cyrtella* on leaves of *Agave*. The prothallus (Fig. 3.3a) enlarges as the algae reproduce by aplanospores and are forced apart by the hyphae. A cortical layer forms and the algae become more or less positioned as a layer below this. Werner calls this structure a primary thallus (Fig. 3.3b), and at this stage the three lichens investigated were still indistinguishable. Next the medulla appears and the algae are restricted to a definite algal layer (Fig. 3.3c). From this secondary thallus there is development toward crustose, foliose or fruticose growth forms. In *X. parietina*, for example, a lower cortex and rhizines are last to be formed.

The secondary thallus is still a submacroscopic structure and considerable growth must occur before a visible lichen thallus can be detected. We do not know how much time elapses between the prothallus stage and visible lichen formation. Data from twig studies where the substrate can be dated suggest a time interval of between one and two years. It takes several months to a year to reach the primary thallus stage in comparable synthesis of lichens from the components in the laboratory.[5] After the new lichen thallus is fully differentiated and attains a diameter of several millimetres, the fungus has sufficient reserve food for the formation of pycnidia and ascocarps. The first-formed reproductive structures are often pycnidia,

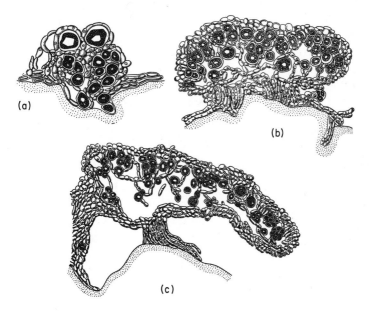

Fig. 3.3 Stages in the development of *Xanthoria parietina* from microscopic sections on *Agave* leaves: (a) prothallus; (b) primary thallus; (c) secondary thallus showing differentiation of lower cortex and rhizines (magnified). (From des Abbayes[1])

and these are formed in a band nearest to the growing edge of the thallus. Apothecia originate farther in toward the centre of the thallus. From studies on dated twigs it is known that ascocarps form two to five years after the thallus becomes large enough to see without magnification.[79,127] Ward observed a regular sequence of lichenization, vigorous vegetative growth, formation of pycnidia, and finally formation of perithecia in *Strigula complanata*,[277] but he could not assign definite time intervals for these stages.

Evidence for sexual reproduction

Except for a few studies that have presented a more or less complete hypothetical life cycle from spore germination to lichenization, there is no real evidence on the effectiveness, need or frequency of sexual reproduction in lichens. A substantial number of crustose species apparently lack any means other than sexual. Foliose and fruticose lichens, on the other hand, can form a variety of vegetative diaspores (Table 3.1, p. 40) and sexual reproduction would appear to be vestigial and functionless. Even some of these, however, may rely on spore dispersal. Degelius[79] found that 11% of the 56 species on *Fraxinus excelsior* twigs and 5 of 22 common oceanic lichens[76] do not seem to possess any obvious means of vegetative dispersal.

There is some indirect evidence for sexual reproduction that, while circumstantial, has significant bearing on this vexing problem. It can be drawn from three sources, monospore cultures, distribution of chemical strains and putative hybridization.

Monospore cultures

There are extremely few studies of mycobionts from monospore cultures. Ahmadjian[7] cultivated hundreds of spores of *Cladonia cristatella* and found a very great and unexpected range of variation in both growth form of the mycelium and in metabolic products, especially pigments. Cultures differed widely in size, shape, production of aerial hyphae and colour. Earlier studies with polyspore cultures had not brought to light this degree of variation. Ahmadjian concluded that sexual processes, fusion of nuclei and recombination of genes could best account for the variability.

Distribution of chemical strains

Formation of chemical strains is assumed to be a reflection of genotypic variation, and patterns of distribution of the strains seem to be as expected as a consequence of spore dispersal. An intensively studied species on pine trees in the eastern United States, *Cetraria ciliaris* (Plate 16B), can be cited as an example.[122] If one collects a large random sample of specimens in a woodlot and determines that 50% of the thalli contain olivetoric acid and 50% alectoronic acid, then specimens *in situ* on trunks and branches will be found to fall into a random arrangement with approximately every other one containing olivetoric acid, the others alectoronic acid. This ratio is unaffected by height of specimens on the trunks and exposure or species of the trees. If the strains reproduced primarily by vegetative means, we would expect clonal, non-random distribution of specimens on the trees. As a matter of fact, *C. ciliaris* has no obvious vegetative propagules, but pycnidia and apothecia are produced in great abundance.

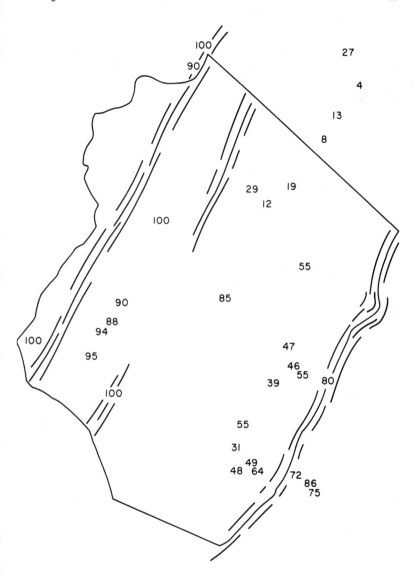

Fig. 3.4 Frequency (in per cent) of specimens of *Cetraria ciliaris* with olivetoric acid (the reciprocal percentage with alectoronic acid) in mass samples randomly taken in 31 stands of *Pinus virginiana* in Pendleton Co., West Virginia, an area of about 1400 square kilometres. Lines represent mountain ridges more than 1000 m high. (Adapted from Hale[122])

On an even larger scale, woodlots separated by several kilometres of farmland have virtually identical ratios of the strains (Fig. 3.4). The ratio changes only when the pine woods are separated by mountain ranges, which in this region run in a general north-east–south-west direction. It appears that each valley or mountain ridge has a uniform population and that intervening mountains act as barriers to mixing of populations with different ratios. This type of distribution pattern could logically be interpreted as a natural consequence of spore dispersal.

Hybridization

Variation in the combination of morphological characters is known to result from interbreeding of sexually reproducing populations. This phenomenon, so widespread in higher plants, could easily account for some of the variability seen in lichens. Certain genera, in particular *Cladonia* and *Usnea*, exhibit a bewildering array of intermediate unclassifiable specimens that seem to be 'hybrids' of two or more clearly defined species.

The only statistical studies of possible introgressive hybridization have been made on intraclonal variation in *C. uncialis* and *C. submitis* in Missouri.[19] Width of the main podetium, podetium tip and the distal region of the podetium, gelification and erosion of the cortex, and the KOH reaction were all studied for a large series of specimens within single clones. The only logical explanation suggested by the authors for the amount of variation encountered was that the original species crossed sexually, and the primary hybrids crossed back to each parent, producing for each species a group of variant forms within which the plant-to-plant variation for one or more of the characters forms a gradient in the direction of the other species. This type of hybridization could be a basis for the notorious variability in *Cladonia*.

4

Physiology and Nutrition

The physiology of lichens can be subdivided into two areas of study: that of the intact composite thallus and that of the fungi and algae isolated in culture. Because of the historical interest in the synthesis of lichens and the admittedly greater ease of controlled experimentation, cultural isolation has been by far the more popular and better known area of study. Yet it is becoming more and more evident that data on the metabolism of the components cannot be extrapolated *in toto* to explain the behaviour of the composite thallus, although such data may be extremely helpful in gauging the physiological requirements of lichens. The thallus must be viewed as a functional unit with properties quite different from those of the components and not a simple additive result of them.

PHYSIOLOGY OF THE COMPOSITE THALLUS[250]

Long-term physiological studies of intact thalli in the laboratory are thwarted by the difficulties of maintaining a symbiotic state, preventing slow dissociation of the components, and eliminating or controlling contamination. Harley and Smith in England made an important breakthrough when they developed and perfected a technique of punching uniform discs 7 mm in diameter from lichen thalli and using replicate samples of 20 or more discs together as experimental material.[130] This technique has reduced the variability of results in using whole thalli that hampered earlier researchers and has made possible meaningful short-term physiological studies. Considerable success has also been achieved recently in controlled-environment whole thallus cultures.[80B, 159A, 159B]

Added to recent successes with radioactive tracers, the pace of research into the metabolism of the composite thallus promises to accelerate.

Water relations

Lichens do not appear to have special organs concerned with the absorption or transpiration of water, nor is there any obvious mechanism for water conservation. Both absorption and loss appear to be entirely physical processes, and in general the water relations of most lichens resemble those of an agar gel. Much of the water in a saturated thallus is held externally to the cytoplasm, partly in the swollen cell walls, partly in the intercellular spaces. The quantity contained in the cytoplasm is not known, but it is likely to be only a small percentage of the total. Different parts of a saturated thallus may contain different amounts of water and different tissues may also vary in maximum water content. Goebel[102] estimated that the central axis of a species of *Usnea* held about one-third of all the water in a saturated thallus. The medulla of *Peltigera polydactyla* contains 25% more water per unit dry weight than the algal layer and upper cortex.[250]

Water content

Measurements of saturated water contents of lichens have been made by numerous workers. For the majority of foliose and fruticose lichens the values lie between 100 and 300% of the dry weight. Ried's[224] work indicates that crustose lichens are comparable in this respect. Much higher values than these have been obtained from the homoiomerous gelatinous lichens, presumably because the thick gelatinous sheath of the blue-green phycobiont is able to hold large amounts of water. Some reported values expressed as percentages of dry weight are: *Collema crispum* 800%, *C. multifidum*, 1400%, *Leptogium lacerum* 830%, and an unnamed *Collema* 3900%.[250] Apart from these species, most lichens have smaller maximum water contents than other kinds of cryptogams, as illustrated by the extensive data on water contents of epiphytic cryptogams assembled by Barkman.[37]

Rate of absorption

Absorption of liquid water by dry thalli of most species is a rapid process. Air-dry thalli become completely saturated after immersion at room temperature in as little as 1 to 2 min and usually not more than 30 minutes. Some crustose lichens are encrusted with non-wettable lichen substances and in extreme cases may not become saturated even after six hours' immersion. Such lichens are assumed to rely mainly on water vapour in nature. While there are wide differences in rates of absorption by various species, the curve of absorption against time is in all cases of the same form as for the imbibition of water by a hydrophilic gel. More than half of the water is absorbed in less than a quarter of the time taken to reach saturation.

When air-dry thalli are placed in a humid atmosphere, their water content rises slowly, over a period of one to nine days, until it reaches a constant equilibrium value. Butin and Ried[250] have shown that at 95% relative humidity the equilibrium water content of 13 foliose and fruticose lichens was 30–50% of the saturated water content. Other workers have found that in an atmosphere of 100% relative humidity 50–75% of the saturation value was reached. The water content of lichens in nature would fall at or below these values, and maximum saturated water content would be achieved only during rain showers and for short periods thereafter.

Water loss

Water loss in lichens resembles that of a hydrophilic gel, and its rate is affected by those factors which affect evaporation. There do not appear to be any special structures or morphological adaptations by which a lichen can control or retard water loss. Rates of loss from intact thalli measured under various conditions of temperature and humidity confirm the common field observation that saturated plants dry out within a few hours in dry weather. The most careful studies have been made by Ried, who showed, as an example, that in a relative humidity of 60% at 20°C the water content of *Umbilicaria polyphylla* fell to 15% in six hours.[224] Under conditions of severe drought, water content of lichens is at very low levels. Field measurements of minimum water contents range from 2.0 to 14.5%.

There are few reliable data on the maximum periods for which various lichens can survive drought. Lange[173] made the only extensive survey and found a range of 8 weeks in *Cladonia impexa* to 62 weeks in *Umbilicaria pustulata*. It is the length rather than intensity of drought that has a greater effect. There is also a broad correlation between habitat and degree of drought resistance, in that lichens from cool moist habitats are less resistant than those from warm dry habitats. Any method of estimating drought resistance should involve prolonged study of the after-effects, otherwise it is difficult to say if the lichen has suffered permanent damage. When a lichen is moistened after drought, the respiration rate may climb rapidly to a value several times higher than normal, followed by a slow decline to the usual rate. By contrast, photosynthesis starts very slowly and rises gradually to the normal rate. Thus, for a time after the end of a drought, there may be a net loss of carbon dioxide from the lichen due to marked stimulation of respiration accompanied by a continued low rate of photosynthesis. These effects can persist for days or even weeks after a long and severe drought.

Photosynthesis

Rate of photosynthesis

Lichens generally have much lower rates of photosynthesis per unit surface area than the leaves of higher plants, although respiration rates of the two types of tissue are frequently comparable. Net assimilation will therefore be very much lower in lichens than in leaves. Ried, working with seven crustose species and five foliose or fruticose species, obtained maximum values of net rate of carbon dioxide assimilation ranging from 0.2 to 3.2 mg carbon dioxide per 50 cm²/h.[250]

Wilhelmsen[283] has provided the most satisfactory explanation for the low photosynthetic rates of lichens. He demonstrated that the chlorophyll content of three species was one-quarter to one-tenth that of leaves. Bednar showed that the algal cells make up only 3–5% of the volume of the thallus of *Peltigera aphthosa*. An additional factor is the opacity of the fungal cortex. Ertl[88] calculated that the cortices of seven species of lichens absorbed 26–43% of the incident light, whereas the epidermis of leaves of three tree species absorbed only 4–13%. To some extent the loss of incident light might be offset by light emitted by fluorescent lichen substances in the cortex.[219] The algal chloroplasts absorb about 25% of the incident light and are not much less efficient than ordinary green leaves.[88]

Photosynthesis and water content

The rate of photosynthesis is closely correlated with the water content of the thallus, although the water content at which the maximum rate is achieved is often somewhat lower than for respiration. The rate usually declines, often sharply, at water contents higher than that at which the maximum rate is reached. This reduction in the saturated thallus is reasonably explained in terms of reduced rates of gaseous diffusion through the thallus to the algal layer. There is also a relationship between the anatomy of the thallus and the level of optimum water content for photosynthesis. The thicker and denser the thallus, the lower the optimum water content. These optima range from 65% in *Umbilicaria pustulata* to 90% of saturation in *Peltigera canina*.[250]

Respiration

Rate of respiration

Rates of respiration for lichens lie in the lower range of values recorded for mature angiosperm leaves. In general, it can be said that the rates for most lichens under optimal conditions at 20°C would fall within the range 0.2–2.0 mg CO_2/g dry wt/h. Knowledge of the respiration rates of thalli of different ages or even different areas of the same thallus is very scanty.

Ried[224] showed that there is no difference in the rates of very young and older thalli of *Umbilicaria pustulata*, although there was a twofold difference in rates of photosynthesis. The medulla of *Peltigera polydactyla* has a lower respiration rate than the overlying algal layer and cortex, but there is no evidence yet of the extent to which the separate algal and fungal components contribute to the total respiration of the thallus in any lichen.[250]

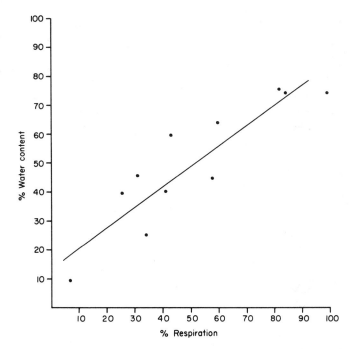

Fig. 4.1   Rate of respiration of *Peltigera canina* at different levels of saturation of the thallus. (Adapted from Smyth[254])

Respiration and water content

There have been many investigations of the effects of changes in water content upon the respiration of lichens, and the same general relationship has been found in all cases. The respiration rate increases with increasing water content in a more or less linear fashion (Fig. 4.1) until it reaches a maximum value. The maximum may be at a water content as low as 40% saturation, but the majority of lichens attain maximum rates in the region of 80 to 95% saturation.[254] Once the maximum rate is achieved, a further increase in water content does not alter respiration rate. The relations between water content and other aspects of respiration, such as respiratory and temperature quotients, are not known.

Temperature

Lichen thalli growing in sunny exposed habitats attain relatively high temperatures. Lange[173] has summarized his own field measurements and those made by earlier investigators which showed that thallus temperatures may commonly reach 50–69°C, values 20–40°C above the ambient temperature. Lange also estimated the limits of heat resistance of more than 40 species of lichens by subjecting air-dry thalli to a range of temperatures for 30-minute periods. The resultant damage was assessed both from measurements of respiration rates and the ability of the algae to resume growth in isolated cultures. The temperature which caused a reduction in the normal respiration rate of air-dry thalli by one-half was taken as an index of the limit of heat resistance, and these indices ranged from 70°C in *Alectoria sarmentosa* to 101°C in *Cladonia pyxidata*. Moist thalli do not differ from other kinds of plant tissue, their limits of heat resistance ranging from 35 to 46°C.

Lichens are remarkably tolerant of very low temperatures. Various species have revived and resumed normal respiration after exposure for several hours to temperatures as low as − 183°C and − 268°C.[183] Although the optimal rate of respiration usually lies between 10 and 20°C, measurable respiration has been recorded at −24°C.[175] Scholander *et al.*[237] compared the respiratory rates of several arctic and tropical lichens, and while arctic species in the families Peltigeraceae and Stictaceae respired at a higher rate than tropical members, most of the lichens had no demonstrable adaptation to counteract climatic influences on metabolic rates.

Seasonal variation

Like other perennial plants, lichens show seasonal variation in some of their physiological characteristics.[249] The experimental evidence, derived mainly from West European lichens, indicates that levels of physiological activity are lower in summer than at other times of the year. The rate of photosynthesis per unit dry weight is generally much higher in winter (December–January) than in summer (May–June) at any given temperature, reflecting perhaps seasonal variation in chlorophyll content. Differences in respiration rate are usually much smaller, so that net carbon assimilation is higher in winter.

In *Peltigera polydactyla* consistent seasonal changes have been observed in the dry weight per unit area of thallus, which rises sharply during late winter and early spring, then falls during the summer. There is a temporary increase in early autumn and a decline at the onset of winter (Fig. 4.2a). Preliminary studies suggest parallel fluctuation in total carbohydrate content. Under laboratory conditions, the rate of absorption of glucose by *P. polydactyla* starts to rise in January and is at a high level by the time the increases in dry weight commence in March; it remains at a high level until June, then falls off during the summer (Fig. 4.2b). Since the main site of glucose absorption is the algal layer, it is assumed that the algae are at a higher level of activity in late winter and early spring than at other times of the year. It is of interest to note that Paulson[211] observed the greatest amount of sporulation of *Trebouxia* in several foliose species to be in spring.

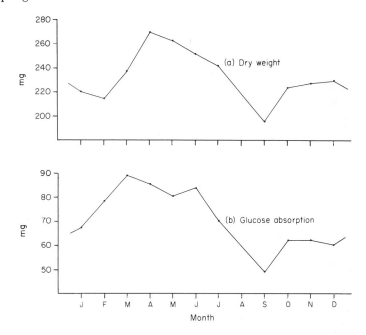

Fig. 4.2 Seasonal variation in samples of 100 discs of *Peltigera polydactyla* collected near Oxford, England : (a) dry weight ; (b) weight of glucose absorbed from a 1% solution in 24 h at 25°C, pH 5.6–5.8. (After Smith[249])

How far seasonal variation in dry weight, glucose absorption, water content and other factors can be translated into seasonal growth rate is questionable. The temperature in summer is optimal, but drought conditions can reduce the levels of assimilation. In winter both lower temperature and reduced light in temperate zones would be unfavourable. As a result, growth of lichens may well be bimodal, occurring primarily in the spring and autumn.

Nutrition and metabolism

The nutrient supply of lichens in nature is dissolved in water that comes into contact with the thallus, and the nutrients are probably absorbed over the whole surface of the thallus. One of the unique attributes of lichens is their efficient mechanism for accumulating a wide range of organic and mineral nutrients, both intra- and extra-cellularly. Laboratory experiments with *Peltigera polydactyla* and other lichens show that the accumulation is a metabolically dependent process because it is sensitive to various biological inhibitors, low temperature and oxygen tension.[250]

Nitrogen

Nitrogen is an essential metabolite for synthesizing proteins in both the alga and fungus. In laboratory experiments Smith[250] has found the nitrogen content of *Peltigera polydactyla* to be 3.6–4.5%. Soluble nitrogen, including ammonia, makes up about one-fourth of the total. When amino acids are supplied externally, there are indications of only a very slow net protein synthesis, and protein breakdown is also small when the thallus is starved. Slow rates of protein turnover might well be a characteristic of all lichens. There are several sources of nitrogen in nature, although we do not yet know what kinds of nitrogen compounds are important for lichens. Nitrates and organic nitrogen, for example, are leached from the tree mantle by rainfall and reach the thallus dissolved in rainwater. Guano accumulated on bird rookeries or perches or in dust on trees near farms are rich sources of organic nitrogen, and species of *Physcia* and *Xanthoria*, among other genera, are frequently restricted to or most numerous in these habitats. This phenomenon has given rise to a large literature on nitrophilous or ornithocoprophilous lichens.[37] However, although these lichens often contain urease enzyme systems, the primary factor that determines their distribution in nature is not definitely known to be nitrogen.

Nitrogen fixation by lichens with blue-green phycobionts is an alternative source of nitrogen and would enable them to inhabit nitrogen-free substrates. It is well established that *Nostoc*-containing lichens fix and accumulate nitrogenous compounds. Even though the nitrogen-fixing bacterium *Azotobacter* has been isolated from a number of these lichens, Scott,[239] using isotopic nitrogen, concluded that it is *Nostoc* that fixes the larger share of nitrogen in gelatinous lichens. Millbank and Kershaw[195A] have also demonstrated nitrogen fixation by *Nostoc* in the cephalodia occurring on the surface of the common *Peltigera aphthosa* (see Plate 8D).

Carbon

The main form of carbon nutrition in lichens is photosynthesis by the algae. Smith and Drew[253] have recently been able to clarify the identity of the photosynthetic products and steps in their translocation. Using discs of *Peltigera polydactyla* in a solution of ^{14}C-sodium bicarbonate, they detected first ^{14}C-glucose, which moved from the algal layer into the medulla within 30 minutes and continued to move at a steady rate equivalent to 40–50% of the total rate of carbon fixation. It was converted at once in the algal layer to ^{14}C-mannitol and at a slower pace into insoluble reserves by the fungus. It is not yet known how far data from *Peltigera polydactyla*, which has the blue-green phycobiont *Nostoc*, can be applied to lichens as a group. Preliminary investigations of *Xanthoria parietina* var. *aureola*, which has a green *Trebouxia* phycobiont, point to ribitol, rather than glucose, as the primary photosynthetic compound which is converted to mannitol.[252]

The extent to which photosynthesis may be supplemented by saprophytic nutrition is not well understood. From the fact that both symbionts in isolated culture can utilize a wide range of organic compounds, it may be assumed that lichens can absorb and utilize some of the simple organic compounds dissolved in liquids passing over the thallus surface or absorbed and transported to the medulla and other layers by rhizoidal hyphae or rhizines.

Mineral elements

Exogenous inorganic metabolites reach the surface of the thallus dissolved in rainwater. Lichens have an unusual ability to absorb ions from their substrates far beyond expectable needs. *Diploschistes scruposus*, a common crustose soil lichen, has a zinc content of 93 400 p.p.m. in Belgium (9.34% of its dry weight), while the soil where it grows contains 10 900 p.p.m.[172] In a large-scale survey of mineral accumulation in Finnish plants, Lounamaa[186] discovered that lichens accumulate more zinc, cadmium, lead and tin than either mosses or higher plants. A Russian group has reported high uptake of both zinc and cerium,[258] and spectrographic analysis of saxicolous lichens in Colorado revealed high concentrations of strontium, titanium, vanadium and yttrium.[177] *Acarospora sinopica*, a saxicolous crustose lichen, contains abnormally high quantities of iron and copper.[176]

Corticolous lichens also selectively accumulate minerals. Mićović and Stefanović[195] made parallel studies of the composition of the ash of oak bark and the attached fruticose lichens. These 'oak mosses' accumulated silicon, phosphorus, magnesium, iron and aluminium to a significant degree. At the same time sulphur, manganese, potassium and sodium were not appreciably accumulated, and the percentage of calcium was less than half that of the bark (Table 4.1).

Accumulation of radioactive elements by lichens is now a well-known phenomenon. Gorham[103] recorded mean (antilog) radioactive count rates

Table 4.1   Mineral composition (%) of oak bark and associated lichens ('oak mosses') in Yugoslavia.[195]

%	Oak bark	'Oak moss'
Moisture	9.15	10.35
Ash	3.95	2.54
SiO_2	4.03	28.42
P_2O_5	1.12	6.37
SO_4	3.20	5.53
CaO	66.58	25.45
MgO	1.10	5.04
Fe_2O_3	0.66	3.16
Al_2O_3	3.79	9.04
Mn_3O_4	0.39	0.36
K_2O	7.60	8.96
Na_2O	1.11	1.50

of 183 in boreal lichens as opposed to 63 for phanerogams at the same localities in Canada. Arctic lichens are especially prone to absorb strontium. High amounts of caesium [137] have been detected in the rock tripe *Umbilicaria mammulata* in Ohio.[57] These radioactive elements were generated by atom-bomb tests and have obviously been washed from the atmosphere and carried to the lichens in rainfall.

The ability of lichens to accumulate and tolerate without apparent harm concentrations of minerals that would be lethal for other plants is truly remarkable, but we have hardly begun to comprehend the mechanisms of ion absorption in the thallus. Lange and Ziegler[176] mention several factors that can be considered, including (1) unspecified cytoplasmic resistance to metallic ions inherent to lichenized fungi and algae, (2) immobilization of the ions within the cytoplasm by means of chelators or other metal-binding substances, and (3) active and passive transport of the ions to regions external to the plasma and cell wall, as localization of insoluble metallic salts on the surface of the thallus. Extracellular deposition can be seen in *Acarospora sinopica* as a thick surface layer of iron salts, but this process has not been found in other lichens that have been investigated.

Beyond the recognition of their unusual capacity for absorbing metallic ions, we know nothing of the mineral requirements of lichens. A considerable amount of indirect evidence leads one to conclude that there are highly specific mineral requirements, but none of these has been proved by experimental study. For example, a large group of lichens in the families

Caloplacaceae, Physciaceae and Pannariaceae have high fidelity for lime-stone and other basic substrates. Perhaps they have a substantial need for calcium or magnesium. *Acarospora sinopica* might require iron since it is restricted to rocks with high iron content. This is a problem for future research.

PHYSIOLOGY OF THE COMPONENTS

Isolation of the component alga and fungus in lichens is comparatively easy and pure cultures can be prepared and studied with the same techniques as used for studying comparable free-living forms. Experiments with the mycobiont present many interesting problems, but slow growth has discouraged investigators. Our knowledge of the physiological requirements of most species is still incomplete, and except for the wide-ranging studies of Tobler,[266] the results of most earlier workers are not trustworthy and need to be repeated. New and more refined cultural methods are constantly superseding older experiments that were carried out on agar or other natural media (see Ahmadjian's book for details[10A]).

Physiology of the phycobiont

Isolation

The phycobiont can be isolated in the following manner.[2] The thallus is washed to remove extraneous debris and, as far as possible, non-lichenized algal epiphytes. Parts of the upper cortex are shaved off with a razor and bits of the algal layer removed, placed in water on a microscope slide, and macerated. Smith[252] has succeeded in making quantitative isolation of algal cells by differential centrifugation of homogenized thalli. Single cells with attached hyphal fragments can be isolated directly with a micropipette and transferred to appropriate media, or the whole slide placed in a moist chamber and alternately rinsed in water and mineral solutions.

Nutrition

Although lichenized algae have been cultured and studied for almost a century, it is still not certain whether they differ significantly in physiology and chemistry from related free-living forms. They tend to grow very slowly in culture, but in many other features resemble other algae.[178] Bednar and Holm-Hansen,[42] however, report that the phycobiont of *Peltigera aphthosa*, *Coccomyxa* sp., excretes 16 times more biotin into the culture medium than free-living *Chlorella pyrenoidosa*. Other workers[96A] have measured more rapid incorporation of supplied $^{14}CO_2$ in *Trebouxia* than in *Chlorella* on a chlorophyll-*a* content per unit dry weight basis, and

Smith[252] discovered further that unusually large amounts of ^{14}C-glucose are excreted by *Nostoc* immediately following isolation and incubation with NaH^{14}CO$_3$, little of the ^{14}C being tied up as insoluble compounds in the cells, but that in a few days excessive excretion ceases and a more normal pattern of ^{14}C fixation resumes. These few experiments lead us to suspect that the fungal hyphae or other factors in lichenization are capable of altering the permeability of the phycobiont so as to promote rapid absorption and excretion.

Trebouxia, the phycobiont most frequently studied experimentally, attains optimal growth on organic substrates and can even grow quite well in the dark if supplied with glucose. Tomaselli[269] found excellent growth on a medium of tryptone, glucose, beef extract and sodium nitrate. Most strains of *Trebouxia* are sensitive to changes in the nitrogen source and can easily be induced to alter their cultural characteristics. Wide ranges of pH and temperature are tolerated,[178] but values for optimum growth are sometimes higher than those required by the associated mycobiont.[138]

Physiology of the mycobiont

Isolation and growth

Spore isolation is preferred and generally used for the mycobiont because there is less danger of contamination. The vegetative thallus can be interlaced by the hyphae of parasymbionts, and one can never be sure that hyphae teased from the medulla represent the true mycobiont. Spores forcibly ejected from apothecia may be caught on the surface of agar-coated slides or inverted Petri dishes and germinated (Plate 11C). If desired, spores can also be isolated individually with a micromanipulator. The mycobiont may be cultivated on solid agar media or in liquid synthetic media, the latter being preferred for controlled experimental studies.

The form and appearance of colonies of different mycobionts and even different isolates of the same mycobiont vary considerably, but in no instance is there any relation in morphology to the composite thallus from which they were isolated. Colonies on solid media tend to be compact and dense (Plate 11A) with little differentiation. Aerial hyphae are characteristic of many species collected in temperate areas (Plate 11B), and Ahmadjian[7] noted a preponderance of smooth rubbery colonies in tropical species. In liquid media some species are slimy (Plate 11E), some flocculent, and some become fragmented into tiny balls (Plate 12).

Growth rates are considerably slower than those for non-lichenized fungi commonly used in mycological laboratories, and the maximum yields and utilization of metabolites are smaller. A rapidly growing mycobiont might reach a diameter of 1–2 cm in a month on nutrient agar, but it is more usual for growth to be only 1–2 mm. Maximum yields in liquid media

Plate 12 Utilization of carbohydrates by the mycobiont of *Sarcogyne similis*. Cultures in 25 cm³ of liquid synthetic medium (carbon at the rate of 10 g/l) harvested after 40 days. Top row (left to right): Dulcitol, D-xylose, D-arabinose, L-arabinose, sorbitol; middle row: cellulose, cellobiose, dextrin, lactose, D-mannitol; bottom row: D-galactose, D-maltose, sucrose, glucose, D-mannose.

are about 100 mg dry weight in 25 cm³ of medium (carbon at 10 g/l),[118] less than one-fourth of the rate one expects with *Penicillium* or *Aspergillus*. Growth peaks, ideally taken from a series of liquid cultures, are reached in as little as three weeks for *Buellia stillingiana* but generally take five to six weeks (Fig. 4.3).[118,138]

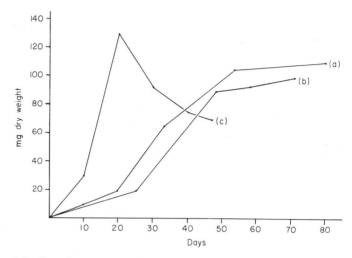

Fig. 4.3 Growth rates of the mycobionts of *Sarcogyne similis* (a), *Acarospora fuscata* (b) and *Buellia stillingiana* (c) in a liquid synthetic medium (carbon at at 10 g/l). (Adapted from Hale[118])

Table 4.2 Utilization of nitrogen sources by the lichen fungus *Cladonia cristatella*. Each reading (mg dry weight of mycelium) represents an average value obtained from growth in six flasks.[7]

Nitrogen source	21 days	45 days	70 days	Initial pH
L-alanine	10.9	47.6	125.5	5.10
L-arginine	13.3	52.5	76.9	4.62
Casein	8.0	42.3	74.9	5.25
Peptone	18.3	40.1	60.9	6.15
Urea	8.4	93.5	11.6	5.35
KNO₃	5.7	25.8	36.9	4.45
NH₄Cl	3.9	17.1	38.6	4.31
NH₄NO₃	10.5	37.9	58.4	4.61
(NH₄)₂SO₄	8.9	44.4	66.2	4.15
No nitrogen	2.1	2.5	3.6	4.45

Nutrition

NITROGEN SOURCES   Utilization of nitrogen by the mycobiont has not been widely investigated and there is such a diversity of results that no general conclusions can be drawn. Ahmadjian[7] carried out comprehensive experiments on utilization for two species: *Cladonia cristatella* (Table 4.2) showed maximum growth with L-alanine, whereas *Acarospora fuscata* gave highest yield on potassium nitrate. *Sarcogyne similis* utilizes ammonium nitrate and asparagine equally well. Tomaselli[269] systematically measured yields on the various amino acids and found that *S. similis* grew best on proline, *Lecidea steriza* on asparagine, although most amino acids gave satisfactory results. Only lysine and phenylalanine failed to support good growth.

Nitrogen source has demonstrable effects on the cultural characteristics of the mycobiont. Conidial formation in *Buellia stillingiana* is inhibited when the nitrogen source is glutamic acid, urea or casein hydrolysate.[127] Production of an unidentified yellow pigment in a species of *Bacidia* from North America is stimulated by ammonium nitrate and glycine but inhibited by ammonium tartrate and glutamic acid.[127] Extremely few studies of this type have been made.

CARBON SOURCES   A large number of different carbon sources have been utilized in culturing mycobionts (Plate 12). Most of the common hexoses provide satisfactory growth, but *Sarcogyne similis* has highest yield on sucrose, *Acarospora fuscata* on mannitol.[118] Ahmadjian[7] observed a marked preference for maltose and lactose in a series of nine crustose lichens from Hawaii. Citrate, acetate, erythritol and trisaccharides are poor sources for the majority of mycobionts so far tested.[120]

As with nitrogen, morphological and physiological characteristics of the mycobiont can be changed or modified by selecting different carbon sources. The hyphae of *Graphina bipartita*, a tropical crustose lichen, have a gelatinous sheath when cultured on lactose or sucrose but none is produced on maltose.[7] This same species synthesizes a bright yellow pigment on lactose but not on sucrose. Dextrose stimulates production of a brown pigment in *A. fuscata*[5] and glucose appears to be necessary for the synthesis of parietin in *Xanthoria parietina*.

VITAMIN REQUIREMENTS   Vitamin deficiencies are probably widespread in lichenized fungi but very few species have been investigated. Studies conducted prior to 1950 suffered from poor or unlucky choices of the mycobiont species and from the use of now unacceptable cultural techniques to prove vitamin deficiency. On the basis of more recent studies, requirements for both thiamine and biotin have been positively identified in *Acarospora fuscata*,[118] *Lecidea decipiens*,[7] *Stereocaulon vulcani*,[7] *Buellia stillingiana*[118] and *Peltigera aphthosa*[42]; *Sarcogyne similis* is deficient only for biotin,[118] and *Collema tenax* for thiamine alone.[138] These and other vitamins are synthesized and excreted by algae, sometimes in large amounts,[42] and made available to the mycobiont in the composite thallus.

pH   The pH of the cultural medium has a very significant effect on the development of mycobiont cultures. Each species has an optimum, usually in the range of pH 5–6; a significantly higher or lower pH will retard growth. *Collema tenax* has highest yield by dry weight at pH 6.2 (Fig. 4.4).

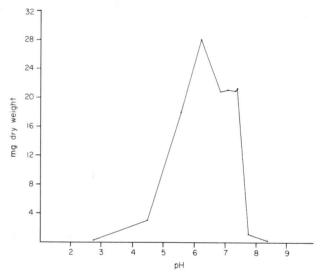

Fig. 4.4 Growth of the mycobiont of *Collema tenax* at different pH levels. (Adapted from Henriksson[138])

Optima for other species are 5.6–6.4 (*Xanthoria parietina*), 5.0 (*Sarcogyne similis*), 4.5 (*Buellia stillingiana*) and 2.0 (*Umbilicaria papulosa*).[120] Effects of pH on the morphology or physiology of the mycobionts have received little attention but are probably as important as those brought about by changing the nitrogen and carbon sources. Conidial formation in *B. stillingiana*, as an example, is suppressed at a pH above 7 or below 5.[127]

TEMPERATURE   The temperature tolerance of mycobionts that have been studied appears to be quite broad. In general, maximum growth occurs between 15° and 20°C, as illustrated by *Collema tenax* (Fig. 4.5). These temperatures are lower than the optima for common *Aspergillus* or *Neurospora* species, which are usually 25° or higher.

LIGHT   As heterotrophic organisms, the mycobionts would not be expected to respond to differences in light intensity or duration. Among the non-lichenized fungi light has a greater effect on reproduction than on vegetative growth. The only reported effect in lichens is a change toward a reddish colour in colonies of *Collema tenax* as length and intensity of light periods increase,[134] reflecting an increase in the production of carotenoids.

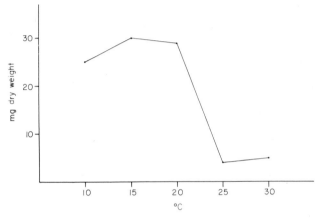

Fig. 4.5   Growth of the mycobiont of *Collema tenax* at different temperatures. (After Henriksson[138])

Products in culture

While the composite thallus is rich in extracellular metabolic products, pigments and various colourless lichen substances, biosynthesis of these *in vivo* by the isolated mycobiont is extremely rare. Production of the orange pigment parietin in cultures of *Xanthoria* is well established, and yellow pulvic anhydride is known in cultures of *Candelariella vitellina*.[6] Hess,[144] on the other hand, cultured seven species of *Cladonia*, *Lecanora* and *Parmelia* on malt-agar and could not detect any depsides or depsidones even after 10 months of growth. Other workers have also failed in attempts to identify lichen substances that were known to occur in the thallus.[7,118] The synthesis of rhodocladonic, usnic and didymic acids by the mycobiont of *C. cristatella*, as reported by Castle and Kubsch,[59] has not been confirmed, and we now believe it to be based on the misidentification of unique substances, including an unrelated red pigment, produced only in culture by the mycobiont.[7] A similar case is *Stereocaulon vulcani*, a fruticose species, which produces atranorin and norstictic acid in abundance in the thallus[7]; the isolated mycobiont synthesizes neither of these, only an unidentified yellow pigment.

Hess did detect two monocyclic phenolic acids, tentatively identified by paper chromatography as orsellinic and haematommic, in several cultures. These acids are precursors of depsides but not of any depsides that are known to occur in the lichens he studied. Hess suggested that algae might intervene to complete the synthesis of depsides, a belief that was held by Zopf[288] but never experimentally proved. Since the fungus *Aspergillus nidulans* is able to synthesize the depsidone nornidulin alone, there is theoretically no reason why lichenized fungi cannot also complete the synthesis of depsides and depsidones under appropriate conditions without algae. Still, the possible role of algae should not be minimized since some are known to produce depside-related tannins.[178] The fungi in the composite thallus depend on the algae for organic substrates and this may be an important factor in the origin and evolution of lichen substances.

The most recent and striking development in this field has been the report on production of lichen substances *in vivo* by Shibata and Kurokawa.[163A] They cultured the mycobiont of *Ramalina crassa* on agar for several months and demonstrated with thin-layer chromatography that the acetone extract of the mycelium contained both salazinic and usnic acids, the components of the symbiont thallus. High culture temperatures ($25°C$) inhibited the production of salazinic acid. The isolated phycobiont, at the same time, did not produce any acids. It is hoped that the success of their investigations will be repeated by other workers.

5

Symbiosis and Synthesis

Symbiosis is the intimate biological union of two different organisms. Lichens, which combine fungi and algae, were the first recognized and are still the best examples of this phenomenon. Mycorrhizal fungi associated with the roots of trees, orchids and other higher plants, and among animals,[98] green algae growing within the coelenterate *Hydra* and eggs of salamanders[101] and frogs also have symbiotic relationships. A unique feature of lichen symbiosis is that the thallus is so perfectly developed and balanced as to behave as a single organism, a new morphological entity with no obvious similarity to either fungi or algae. In fact, no one before 1866 had comprehended the dual nature of lichens. Tulasne identified the hyphal threads with those of fungi in 1852, but the algae, called gonidia, were thought to have an ontogenetic origin as buds from tips of hyphae, which on breaking free developed independently.

De Bary[39] in 1866, followed by Schwendener[238] in 1867, finally came to the realization that the gonidia are no different from free-living algae and that lichens are a union of two unrelated organisms. Many leading workers of the day, however, never agreed with this conclusion but insisted that lichens were autonomous organisms, coequal with fungi and mosses. The opposition from systematic botanists was especially bitter, perhaps because they feared that lichens would lose their special identity when integrated with other fungi. As late as 1930, Elfving[86] tried to demonstrate that the 'algae' were budded from hyphae in the gonidial layer, and Schmidt[236] in 1953 published what will probably be the last serious attempt to defend this discredited argument.

As soon as the dual hypothesis of lichens was announced, lichenologists sought urgently to explain the relationship between the fungus and alga.

After almost 100 years of polemics, definitions and theories, we are still unable to understand completely the mechanisms of lichen symbiosis. The following brief summaries of the terms often seen in textbooks may help clarify the basic concepts.

CONCEPTS OF SYMBIOSIS

Parasitism. A parasite is a heterotrophic organism that is able to establish a nutritional relationship with a host, a relationship that is favourable to itself but usually unfavourable to the host.[38] A necrophytic parasite attacks and eventually destroys its host, often by haustorial penetration of the protoplast. Schwendener envisioned the fungus in lichens to be parasitic on the enslaved algae and described this condition as helotism. There are undoubted examples of intracellular haustoria in some crustose lichens. The permanency of the lichen thallus seems to refute a simple parasitic relationship, but this does not exclude the possibility that the algae, though parasitized, are able to reproduce faster than they are killed by the fungus.

Algal Parasitism. Moreau, the renowned French mycologist, considered the lichen association to be algal parasitism, a chronic pathological state in which the hyphae are stimulated to form a large vegetative mass on contacting algae, the thallus manifestly being equivalent to a gall or cecidium.[200] This curious concept has been given little support by lichenologists.

Endosaprophytism. The Russian lichenologist Elenkin[85] subscribed in part to the idea of parasitism but modified it to include a certain degree of saprophytism, a plausible concept since most fungi have saprophytic nutrition. The hyphae would be able to live off dead algal cells that lie in a 'necrotic layer'. It is doubtful, however, that the comparatively few dead algal cells seen in the algal layer of most lichens could provide adequate nutrition for the fungus, although a limited amount of saprophytism cannot be ruled out.

Mutualism. This term describes a mutually beneficial relationship where one or both components may be dependent on the association for survival. Reinke in 1872 had called this union consortism, and a lichen is commonly referred to as a consortium. Among the fungi a kind of mutualistic symbiosis is postulated for the biotrophic parasites,[38] which attack a host but soon establish with it a close nutritional relationship, absorbing nutrients by means of intracellular haustoria or appressoria without apparent injury to the cells. An important feature of biotrophism is that one member of the pair cannot grow alone, but is dependent on the other not only for nutrients but also for specific growth factors.

Polysymbiosis. The discovery of nitrogen fixation by various lichens has given birth to the concept of polysymbiosis. The main tenet of this concept is that some lichens are a combination of a fungus, an alga and a nitrogen-fixing bacterium, usually *Azotobacter*, all acting in unison as a polysymbiont. While *Azotobacter* has been isolated from some lichens, Scott[239] is not convinced that polysymbiosis is a realistic theory. It must rest on proof that nitrogen is fixed within the thallus by *Azotobacter* and the products of fixation utilized by the alga or fungus. While the phycobiont *Nostoc* does fix appreciable amounts of nitrogen, there is no evidence at present that any fixation attributable to *Azotobacter* is of significance to symbiosis.

Parasymbiosis. Parasymbionts, Discomycetes or Pyrenomycetes that infest lichen thalli (Plate 9F), are rarely found apart from lichens and are closely related to saprophytic species growing on mosses and marine algae. Their hyphae invade the whole thallus, even the ascocarps, but are unable to form a discrete vegetative thallus. Parasymbionts derive nutrients from the algae and maintain a balanced state of symbiosis with them and with the composite thallus. They may be unlichenizable fungi or degenerate, previously lichenizable fungi that carry on a vicarious symbiotic existence. Their physiological relationship to the principal symbionts is a problem that deserves further study.

All proposed concepts of the relation between the fungus and alga in lichens contain some elements of truth, but no one term can, in fact, embrace the complex physiological activities that make lichen symbiosis such a highly successful venture in the evolution of algae and fungi. We know far too little to make a choice among the various theories. The fungi could engage in both parasitic and saprophytic nutrition; in gelatinous lichens they seem to have a commensal relationship with *Nostoc*. A balance between the components, so self-evident in the continued existence and longevity of the thallus, could scarcely be maintained without some degree of mutualism.

Is lichen symbiosis so highly evolved that in effect a new organism is created? In 1907 Famintzin[90] spoke of lichens as new physiological entities that deserve a separate niche in the plant kingdom. More recently, Culberson[69] expressed opposition to recent changes in the *International Code of Nomenclature* that seem to equate lichens with fungi and imply that contributions of fungi and algae are additive. Lichens are undeniably more than a sum of their parts, for lichenization is accompanied by structural modifications (e.g. thalloid exciple, soredia) new to the plant kingdom and physiological activities (production of lichen acids) different from those of either component.

PHYSIOLOGICAL RELATIONSHIPS OF THE SYMBIONTS

Dialectic arguments of symbiosis in lichens are no substitute for experimental proof. It is all too evident that an understanding of the physiological interactions between fungi and algae is the only answer to this difficult problem. De Bary long ago theorized that the symbionts excrete and benefit from an exchange of metabolites, the alga receiving minerals, water and nitrogen from the fungus, the fungus receiving carbohydrates from the alga. This idea was reinforced by Quispel's pioneering work in 1945[218] with weakly lichenized 'algal covers', in which there seemed to be an exchange of vitamins and unidentified growth substances. Although his results are not directly relevant to the more advanced lichens, the experiments serve as a model for the type of research that needs to be done.

Relationships of the alga

The advantages that accrue to the algae seem fairly obvious. They gain mechanical protection from the elements by being tightly enveloped by hyphae and should also benefit from somewhat improved water relations and resistance to desiccation, although there is no proof that free-living algae cannot withstand drought as well as lichenized algae. Indeed, free-living soil algae can withstand dry periods of up to 70 years. The opaque hyphae protect the algae from high light intensities, since there is a considerable reduction of light within the thallus.[88] Ahmadjian[5] found that *Trebouxia* has poorest growth in highly illuminated cultures (4300 lm/m^2), as had previous workers, but the same phycobiont prospered in culture when surrounded by hyphae. It may then be that it is not a case of *Trebouxia* being able to adapt to low light intensity, it may actually be dependent on it.[4] Photophilic algae as *Pleurococcus*, when lichenized, would be inhibited by reduced light and selectively killed off by the fungus.

Aside from mechanical protection, the algae may improve their nutritional regimen by an exchange of metabolites with the fungus, as Quispel proposed. Being heterotrophic organisms, the fungi would hardly seem overmuch concerned with the welfare of the algae, but they may be able to excrete as yet unidentified substances that promote photosynthesis in algae. The algae cannot help but benefit from the minerals that are accumulated in the lichen thallus in high concentrations.

The primary disadvantage for algae in symbiosis is assumed to be parasitism by fungal haustoria. Haustoria in non-lichenized fungi, it is true, function either to penetrate and kill living host cells or to absorb nutrients from them by extracellular digestion. There is actually little evidence that haustoria normally perform either of these functions in lichens or, if functional, are there any data on the number of algal cells affected. Smith[251] argues that haustoria may represent a stage in the evolution of lichens from a condition where the ancestral fungi and algae lived commensally and now serve largely to increase the area of contact between the symbionts. Recent physiological experiments, while incomplete, seem to give evidence of a more detrimental effect than damage by haustoria; namely, that the permeability of the algal cell wall is somehow altered by the fungus, allowing nutrients to leak out. Symbiotic algae, when first isolated, may excrete abnormally large quantities of glucose and vitamins, a characteristic that is gradually lost in culture.[252]

Relationships of the fungus

The lichen fungus must have recourse to organic nutrients synthesized by the algae since they are the only component in the thallus capable of elaborating these. Even when lichens grow on organic substrates, humus or bark, exogenous metabolites are probably transported to the thallus only to a very limited extent. Henriksson[136] has demonstrated abundant secretion of polypeptides and polysaccharides by symbiotic *Nostoc*, although polysaccharides appear to be a poor source of carbon for the fungus.[138] Vitamin deficiencies seem fairly widespread in fungi,[118] and the fungi can be expected to benefit from vitamins synthesized by algae. Where the phycobiont is *Nostoc*, nitrogen fixation should result in additional sources of organic nitrogen to the fungus.

Disadvantages for the fungus in symbiosis are few. There may be possible antagonistic effects on the hyphae by algae,[135] but these are probably minor in comparison with the detrimental activities of fungi on algae. The lichenized fungus achieves greater longevity than other fungi, though the price in terms of slow growth is high. The delicate symbiotic balance is relatively easily upset by slight changes in the environment, exposing the fungi more acutely to the whims of nature. The fungus is wholly dependent on symbiosis with algae to form sexual fructifications, although many lichens no longer have need for sexual reproduction.

Evolutionary significance of symbiosis

No amount of discussion or experimentation can fully communicate the enormous success of symbiosis in nature. That so many inoperculate fungi should have adapted themselves to symbiosis is indisputable evidence of this. Most lichenized fungi could surely not have survived to this day without the co-operation of algae and indeed, the most common phycobiont, *Trebouxia*, is so rare as a free-living alga that it too appears to be dependent on symbiosis for survival. Perhaps it has evolved from a closely related free-living genus such as *Chlorococcum* through natural selection caused by lichenization.[10] At some early stage in evolution, both the lichenized fungi and *Trebouxia* may have been faced with near extinction in competition with non-lichenized fungi and algae. Through lichenization they have passed from a very precarious existence to their present numerical superiority over other plants in pioneer habitats.

SYNTHESIS OF LICHENS

Schwendener's reports on the nature of lichens stimulated not only theoretical discussions of symbiosis but also experimental creation of lichen thalli designed to support or disprove the idea of dualism. After all, if lichens are dual, it is only natural that a thallus could be reconstituted by combining the isolated components (Fig. 5.1). Considering the comparative ease of separating the components, however, one is not prepared to accept the fact that complete synthesis of a lichen starting with the components and ending with a fertile thallus has never been accomplished.

Since the late 1870s there have been many attempts to reconstitute thalli, and at least superficially a few seem to have been successful. Ahmadjian,[5] however, reviewed these experiments and concluded that only three workers, Stahl, Bonnier[47] and Thomas, seem to have succeeded in achieving a true synthesis, and the validity of Stahl's and Bonnier's work is at best doubtful. All too often contaminated or otherwise imperfect cultures gave the mistaken illusion of success. It was not until 1939 that Thomas[262] presented convincing experimental evidence of synthesis. He concluded that lichenization cannot occur under conditions that favour the growth of one component over the other. Excessive moisture, for example, will often stimulate the overgrowth of the alga. The use of media rich in organic nutrients will actually induce dissociation and the fungus and alga merely grow side by side.

Ahmadjian has been able to achieve a considerable measure of success in reconstituting lichens. His experiments may be described briefly as follows.[5] The mycobiont and phycobiont of the common crustose *Acarospora fuscata* were isolated and grown in pure culture. Synthesis was attempted in 250 cm^3 flasks with purified agar and no other added nutrients. The fragmented mycobiont culture was mixed with the algae in a Waring blender so as to form a suspension which was inoculated on the agar surface. Ten days after inoculation the hyphae had begun to encircle the algae and appressorial appendages were noted. After 15 days the alga seemed to deteriorate, turning yellowish and granulated. The fungus, after recovering from an initial period of disintegration, appeared healthy, and by the end of 30 days had completely enclosed the algae with pseudo-parenchymatous tissue. Algal colonies embedded in fungal tissue proliferated into compact masses of bright green cells. After nine months, structures unmistakably resembling cortex and medulla were visible under the microscope, although it was obvious that months or even years would be needed to form a thallus equal in size to the naturally occurring thallus.

COMPOSITE THALLUS

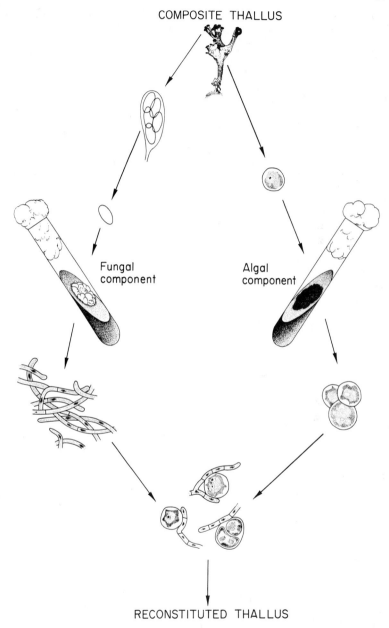

Fungal
component

Algal
component

RECONSTITUTED THALLUS

Fig. 5.1 Steps in the experimental reconstitution of *Cladonia grayi*. (Drawings by N. Halliday)

In another series of experiments, the components of *Cladonia cristatella* were inoculated in flasks containing a wood substrate partially embedded in nutrient agar.[9] The cultures were allowed to dry out over a period of several months. At the conclusion of the tests, the cultures contained squamules with structural organization similar to that found in the naturally occurring lichen as well as soredia-like structures which are not characteristic of *C. cristatella*. At the same time pycnidia and podetia tipped with ascogonia and an undifferentiated hymenium had formed. Drying prevented further development of the cultures.

Ahmadjian and Heikkilä[10B] had success recently in duplicating the classic synthesis of *Endocarpon pusillum*, a small squamulose soil-inhabiting lichen. It contains algae in the hymenium of the fruiting bodies as well as in the thallus. Stahl in 1877 had used the fungal spores with attached algae to initiate the stages of synthesis. Ahmadjian, however, recombined algae-free spores and the isolated alga in soil cultures; a fertile thallus developed after several months. Spores from the newly formed fruiting bodies of the artificially reconstituted thalli were then used to initiate another synthesis with the phycobiont, and they too produced new colonies.

The laboratory conditions necessary for synthesis of lichens involve appropriate alternation of moisture and nutrients. Techniques for routine synthesis will undoubtedly be perfected within the next decade. Because of the overall slow growth of lichenized fungi, however, we should not expect any dramatic breakthrough. Improved methods of synthesis will bring within the realm of possibility the solution to many unanswered questions on the specificity of algae and fungi in lichens. For example, various combinations of the mycobiont with unrelated algae could be made to evaluate the role of algae in determining the identity of a lichen. The effects of combining two or more different mycobionts could be profitably studied. Eventually it might be possible to create new 'species' of lichens.

6

Growth and Longevity

Lichens have a basic centrifugal pattern of growth, and unless the substrate or other plants interfere, the thallus will grow outward uniformly at the margins and form an orbicular colony. Fruticose lichens continually grow in length. In early stages most growth activity is funnelled into increasing the surface area of the thallus. Beyond a certain point lateral transport of nutrients between the margin and centre of the thallus is hampered and the older parts thicken somewhat or become folded, produce isidia, soredia and other vegetative structures, or divert energy into the formation of pycnidia and ascocarps. The older parts seem, however, to remain in a functional state throughout the life span of the lichen.

GROWTH RATES

It is axiomatic that lichens grow slowly. There are comparison photographs made over periods of several decades that seem to indicate no increase in thallus size for both crustose and foliose species. Lichens with diameters of 30–40 cm that grow in the Arctic are reputed to be of the order of several thousand years old.[44] But it should not be forgotten that only direct and continuous observation can establish the true age of a lichen thallus. It is, on the other hand, incorrect to say that little is known about the growth rates of lichens. There are many published records which, taken as a whole, provide a sound basis for generalizations. Some typical values for average annual radial rate (mm), mostly for temperate species, are listed in Table 6.1.

Table 6.1   Rates of growth.

Species	Average annual radial rate (mm)
Foliose species	
Parmelia caperata	4.0[127]
P. conspersa	1.6,[119] 5.3,[213] 7.6[127]
P. olivacea	0.7[182]
P. saxatilis	0.5–4.0[97]
P. sulcata	1.72,[52] 2.22,[79] 1.2,[127] 1.2[182]
Peltigera canina	3.0–7.0[97]
P. rufescens	25.0–27.0[97]
Physcia stellaris	1.1[127]
Umbilicaria cylindrica	0.01–0.04[97]
Xanthoria parietina	2.5[79]
Fruticose species	
Cladonia alpestris	3.4[241]
C. coccifera	1.6–2.0[97]
C. rangiferina	2.0–5.0,[97] 4.1[241]
Evernia prunastri	2.0[79]
Ramalina reticulata	11.0–90.0[142] (7 months)
Crustose species	
Diploschistes scruposus	2.0,[97] 2.0–3.0,[44] 0.44[119]
Lecanora alphoplaca	0.67–1.40[97]
Lecidea coarctata	1.4[119]
L. elaeochroma	1.25[79]
Rhizocarpon geographicum	1.0[44]
R. oreites	1.0–1.5[44]
Rinodina oreina	0.57[119]

The rather wide variation in values reported by different investigators is a reflection of the range both in growth that can be expected and in accuracy between different methods of measuring growth. Furthermore, it is evident that the growth of any one species varies according to the locality. *Parmelia conspersa* in Connecticut, where winters are long and cold and the summers cool, grows at an average rate of 1.6 mm per year (1949–64),[119] but this same species grows significantly faster in Tennessee, 5.3 mm per year (1958–61),[213] where the winters are mild and the summers hot.

Measurement of growth

The methods of measuring growth rate have unfortunately not been standardized. It is, in fact, not always apparent how earlier workers determined growth, whether, for example, by measuring individual lobes, by compiling averages of several lobes or by taking the diameter of a whole thallus. A recent technique, especially suitable for rapid field study, is to trace the outline of a thallus on transparent plastic sheets and, after some months or years,[52,119,228-9] retrace the same thallus, positioning the sheet by means of permanent orientation markers (Fig. 6.1). This is at best a rather crude technique for short-term measurements because errors of up to a millimetre are unavoidable when marking the plastic, but over a period of several years these errors diminish relative to total growth and average out. On the other hand, if tracings are made at too infrequent

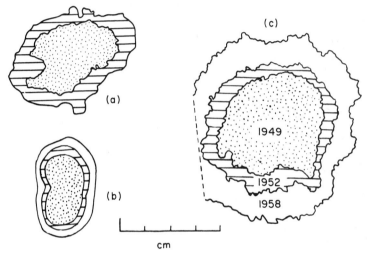

Fig. 6.1 Lichen thalli traced on plastic sheets : (a) *Graphis scripta* (crustose) in Poland, 1954 and 1960 (after Rydzak[229]) ; (b) *Rinodina oreina* (crustose) and (c) *Parmelia conspersa* (foliose) in Connecticut, 1949, 1952 and 1958 (from Hale[119]).

intervals, it is often difficult to identify and follow individual thalli and lobes. Finally, in the case of corticolous species, horizontal expansion of the trunk may stretch lichens 1% or more each year; vertical dimensions are unchanged. A correction factor must be calculated from changes in the position of orientation points.

Close-up photography of thalli is probably the most desirable method of measuring growth rate, since the photographs can be kept for permanent reference. Measurements of individual lobes accurate to 0·1 mm are possible by enlarging the photographs (Plate 13). By far the most complete study of this type was recently published by Frey[97] from quadrats set up in Switzerland and photographed at intervals of 4 to 11 years from 1922 to 1956. He was able to show both growth of lobes and whole thalli and succession and changes in lichen communities.

Plate 13 (A) Thalli of *Parmelia caperata* (lower third), *P. conspersa* (left centre), and *Physcia millegrana*. (B) Same quadrat four months earlier. Black lines are for reference and orientation. Taken near Washington, D.C.

Growth rates may also be expressed as increment in surface area. More or less rounded thalli are traced on plastic sheets and the area measured with a planimeter. The difference between this area and that obtained by a second tracing at a later date gives a value for total growth.[229] If the radius is calculated at the same time by the formula, radius $= \sqrt{\text{area}} \times 0.5642$, a useful value of average radial growth can be obtained.[119] Variation in lobe-by-lobe and absolute linear rate is not given by this method and must be measured separately from the plastic sheets. Rydzak[229] interpreted per cent increase of surface area as an expression of growth rate, but in general the rate of increase is a function of the original size of the thallus, a mathematical rather than biological phenomenon. Therefore thalli of approximately equal size should be used in comparative studies on growth rate.

An indirect method of assessing growth is to measure thalli on dated substrates. This is a convenient short cut because it does not compel one to wait patiently for years as the lichen slowly grows. Many man-made objects, fence posts, sign boards, shingles, bridge abutments and above all dated monuments and tombstones, are invaded by lichens. If the date of origin of the substrate is known or calculable, an approximation of minimal growth rate can be reached by dividing the years of exposure into the thallus diameter. The basic difficulty here is that the dates of origin of the substratum and invasion by the lichens do not necessarily coincide. There is a variable lag period, one which can be estimated accurately for dated twig internodes at three to five years but which is very difficult to state for other kinds of substrates. Growth rates computed from twig analysis, which take a lag period into account, are generally comparable to growth data calculated from direct and continuous observation of the same lichens.[79]

Conversely, the size of lichen thalli can be used to date a substrate of unknown age. Beschel[44] has intensively studied the dating of glacial moraines with lichen thalli, a method that he calls lichenometry. The diameters of normal orbicular thalli on dated moraines are measured and graphs of age versus diameter plotted as a reference standard. By extrapolation, the diameters of the same species, most often *Rhizocarpon geographicum*, on undated moraines can supply a more or less reliable approximation of the date of formation for the moraine. Beschel has dated moraines as far back as 1000 or more years. This method appears to have considerable merit in glacial geology. Follmann[95] used the same technique, in conjunction with photographs taken at an interval of 47 years, to assign an age of about 400 years to the stone images on Easter Island.

Seasonal variation

Lichen growth rates vary according to the season. Both Phillips[213] and Rydzak[229] have shown that summer rates are significantly higher than winter rates and, in fact, no growth at all may occur in cold winters. A month-to-month study of *Parmelia baltimorensis* in Maryland showed an expectable pattern of a low growth rate in winter and a spurt in the cool, moist late spring and early summer months (Fig. 6.2).[127C] Year-to-year

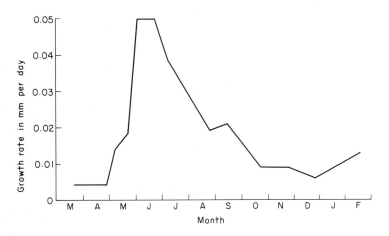

Fig. 6.2 Average daily growth rates of a single lobe of *Parmelia baltimorensis* measured over a period of 12 months.[127C]

variation correlated with broader climatic fluctuations has been reported by Brodo[53] for common foliose lichens in New York.

LONGEVITY

In spite of reports on their persistence and agelessness, lichens surely have a maximum thallus size and finite life spans. The scant information on this subject seems to conflict. *Cladonia rangiferina* and related reindeer mosses, for example, appear to have a normal sigmoid pattern that is divisible into three phases: initial, maturation and degenerative.[241] In the initial period the podetia grow steadily and branch. During maturation the base of the podetium decomposes and the rate of decomposition equals the rate of new apical growth. In the final degenerative period the podetia decompose faster than they grow and eventually perish. The overall growth rate as measured by distance between branching axils is estimated to be 4.1 mm per year. The total growth period is usually from 10 to 25 years, but may be considerably longer.

A sigmoid pattern of growth is characteristic of *Parmelia conspersa*, a common saxicolous lichen. The thallus grows relatively slowly until a diameter of 1.0–1.5 cm is reached, the actual rate being about 0.8 mm per year in Connecticut[119] and about 2.5 mm in Tennessee,[213] approximately half that of larger thalli. Accurately measured monthly rates of juvenile thalli in Virginia are also significantly slower than those for mature thalli in the same rock quadrats. Growth of plants more than about 1.5 cm in diameter continues more rapidly in a linear fashion, but as the thalli approach 10–12 cm in diameter, there is a noticeable slowing. Very frequently the central part of the thallus crumbles and falls away. Remaining marginal lobes may regenerate but the identity of the original orbicular colony is lost. The total life span of this lichen in Connecticut can be estimated to be about 39 years, during 16 of which measurements have been made. Linkola[182] has arrived at similar estimates of life spans, some ranging as high as 80 years, by extrapolation from observed growth rates.

Other studies have discovered different patterns of growth. Beschel[43] measured growth from dated grave markers and convincingly showed that after a very brief lag period during establishment, growth is most rapid in juvenile thalli, reaching a peak between four and eight years, levelling off, and continuing at a linear rate through maturity. Nienburg's studies[206] on *Pseudevernia furfuracea* on dated twigs seem to indicate the same pattern with rapid initial growth during the first six to eight years, then a drop off.

More definitive conclusions on the patterns of growth, seasonal variation and life spans of lichens cannot be drawn at this time. Is early growth logarithmic? How soon is a maximum thallus diameter reached and what limits it? Do slow-growing Arctic lichens differ in pattern from more rapidly growing species in temperate areas? Well-planned investigations specifically aimed at answering these questions have been started only in the last 15 years, and life spans seem to average at least 30 to 50 years. Truly fundamental studies of lichen growth, therefore, cannot be completed for some years yet, if we assume that continued study of individual thalli by direct observation is the most reliable method of assessing growth.

EFFECTS OF ATMOSPHERIC POLLUTION

Lichens are extremely sensitive to man-made atmospheric pollutants, and they are the first plants to disappear in cities. The progressive loss of lichens in major European cities was first noticed almost 100 years ago; the central areas of Bonn, Helsinki, Stockholm, Paris and London contain no lichens. Surrounding the central 'biological desert' is a zone of depauperate *Physcia* or *Xanthoria* species along with various impoverished crustose lichens. Beyond that in the suburbs is a transition to the native lichen flora where large fruticose and foliose lichens are conspicuous.[37, 100B] The effect is even more spectacular around smelting or sintering plants which are more concentrated pollution sources than cities.[176B, 220A] Sharply defined biological zones are formed within a few years. A recently published book[90A] devoted entirely to air pollution and lichens reviews this aspect and many other topics.

Some lichens nevertheless show surprising resistance to pollution. Crustose *Lecanora conizaeoides* is apparently confined to industrial areas in Great Britain[154] and is not found outside cities. *L. hageni* and species of *Lepraria* are also quite tolerant. Soil-inhabiting lichens on the whole are less susceptible to damage than corticolous species. *Stereocaulon pileatum* is reported as a soil lichen from several urban localities in England and a crustose lichen actually grows on flowerpots in Paris greenhouses.[71]

Causes of lichen death

When a healthy lichen is transplanted from its natural environment to a city with a polluted atmosphere, the green algal layer fades within a few months and the lichen dies.[51] There are of course many complex pollutants in cities, but one, sulphur dioxide, is usually considered to be a primary causative agent. Skye[247] in Sweden correlated the distribution of lichen-free trees with the concentration patterns of SO_2-laden fumes which were being discharged from a shale-oil rendering plant. In general, lichens growing within a radius of 6–8 km of the plant had disappeared within 10 years. After this plant was closed in 1966, Skye reinvestigated the lichen flora and found that some of the lichens were recovering and showing new growth.[247A] Brightman[48] explained the occurrence of rather rich lichen growth only 13 km from the centre of London as being related to the power of the asbestos shingle substrate to absorb and neutralize SO_2.

The deleterious effect of SO_2 has actually been proved in laboratory experiments. Rao and LeBlanc[220] subjected several common corticolous lichens in Canada to controlled amounts of SO_2 at various humidities and found that the algal components were extremely sensitive to injury by sulphurous acid and sulphates. Chlorophyll-*a* was converted to phaeophytin-*a* by the loss of magnesium. At very high concentrations the direct action of sulphurous acid caused irreversible damage to the cells. Since lichens have relatively less chlorophyll than other plants, even small amounts of SO_2 could wreak havoc on the delicate balance between the alga and fungus. The inability of lichens to excrete toxic elements, coupled with their efficient mechanisms for accumulating them, also enhances their susceptibility to atmospheric pollution.[250]

The effects of SO_2 are felt over very broad geographic areas. Another important pollutant, fluorine, has a more limited but equally destructive effect. Gilbert,[100A] for example, observed a pronounced lichen desert extending out several kilometres from an aluminium smelting plant at Fort William, Scotland, that could be correlated with the accumulation of various fluorides by the lichens.

Rydzak[227] has made especially detailed studies of lichens in Polish cities and presents convincing data that the low humidity and higher temperatures of cities alone can explain the loss of lichens. Desiccation and lack of dew formation are characteristic of the urban microclimate, and lichens are subjected to abnormally rapid wetting and drying cycles. While we should not minimize the possible effect of desiccation, controlled experiments[212] designed to assess the comparative effects of desiccation and SO_2 point to SO_2 as being measurably more damaging. Thalli exposed to SO_2 are thinner and discoloured and the rate of photosynthesis is reduced, as expected from the known action of SO_2 on chlorophyll. Lichens merely kept dry in an SO_2-free atmosphere can still be revived after as long as three years and resume normal photosynthesis.

The sensitivity of lichens to atmospheric pollution has led to their successful use as a kind of bioassay of pollution levels.[217A] Indicator species are assigned numerical values corresponding to their known resistance, as Hawksworth and Rose did for British lichens,[131C] and by the use of appropriate formulae an index of pollution can be computed.[80A] As concern about the harmful effects of pollution in urban areas grows, the value of lichens as qualitative and even quantitative indicators will become more appreciated.

Pollutants and desiccation are not the only factors that correlate with the loss of lichens. The continual destruction of native trees in cities and replanting with lichen-free nursery stock effectively reduce the supply of vegetative propagules.[37] This situation could also account for the noticeable reduction of the lichen flora in heavily farmed areas such as the Netherlands, southern England or east-central United States, where old herbarium records of showy fruticose and foliose lichens made before the primary forests were cut down simply cannot be duplicated in secondary or regenerated forests. Atmospheric pollution is not a factor here at all nor could desiccation account for the loss.

A final example of pollution is that resulting from oil spills by ocean-going tankers. Ranwell[218A] surveyed damage from severe crude oil pollution along 320 km of the Channel coast in Cornwall and Brittany. While lichens in the midlittoral zone were smothered by up to 2 cm or more of the oil, most destruction occurred with the emulsifiers used to break up and remove oil deposits. The long-term damage that results will obviously affect the littoral ecosystem for years to come since many invertebrate animals depend on lichens as a food source.

It is not only man who is modifying our environment to the detriment of lichens. More and more studies are showing that animals have a significant effect on lichen survival which, coupled with pollution and tree removal, can explain the steady decline in lichen vegetation. Snails and slugs are well known as lichen predators.[62A] A particularly destructive invasion by springtails has been noted near Washington, D.C.,[127D] and various other insects and mites have been implicated in other areas.

EFFECTS OF IONIZING RADIATION

Another type of atmospheric pollution that is a potential hazard in some areas is atomic radiation. Here the lichens fare much better than other

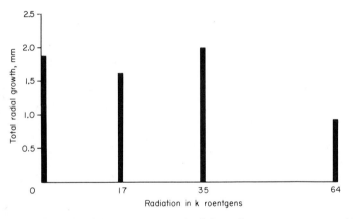

Fig. 6.3 Effect of radiation on the growth of *Parmelia conspersa* measured for a six-month period between March and September 1965. Caesium*[137]* gamma radiation administered at the rate of 130 roentgens/h for 22 h a day : 17 kR for 6 days, 35 kR for 11 days, and 64 kR for 24 days. Reduction in growth at 64 kR statistically significant at the 99% level. (After Jones[155])

plants. Brodo[52] has found in preliminary investigations that lichens are unusually resistant to levels of radiation that completely kill all other plants within a matter of months. Jones[155] conducted controlled experiments on radiation effects on *Parmelia conspersa* over a period of six months in Georgia. Statistically significant reduction in growth rate occurred only at levels of 64 000 roentgens and above (Fig. 6.3). Several factors may account for this resistance. The vegetative lichen thallus is dormant much of the time when dry and the meristematic area is diffuse. In addition the nuclear volume is very low and the chromosomal structure rather simple.

7

Ecology and Succession

Phytosociology is probably the most thoroughly investigated aspect of lichen biology. Communities in temperate regions have been surveyed by numerous workers and in Europe many have been formally named according to various phytosociological systems. Barkman[37] has published an excellent text on cryptogamic epiphytes that draws together all data up to 1958 and forms an indispensable reference for students interested in the phytosociology of lichens.

SUCCESSIONAL PATTERNS

Lichens have a well-deserved reputation as pioneers in plant succession. They occur in virtually every pioneer terrestrial habitat from arctic and antarctic to tropical areas, where they are able to form long-lived and stable communities. It is their adaptation to existence in xeric environments that has enabled the majority of lichens to dominate in habitats where competition from other plants is very slight. This adaptability is closely correlated with the photophilous (light-loving) characteristics of most lichens.

Lichen succession is largely directional, and changes in the environment determine the ultimate fate of the communities. Stages in succession can be arrested if the environment either remains unchanged or is not subject to change, conditions that are most easily met in the Arctic and Antarctic, deserts, rock outcrops in temperate areas where growth of trees is prevented, and rocks along rivers, lakes or the oceans. In these habitats lichen communities can last for centuries. If, however, they become established in habitats where there is subsequent shading by shrub or tree canopies or an accumulation of humus and leaf litter or where the substrate is unstable, the communities are eventually replaced by mesophytic lichens, bryophytes and higher plants. This pattern of succession is typical of forested areas, lava flows that become forested, abandoned fields and similar habitats.

Lichen communities may be classified conveniently according to their substrate: tree bark, wood, rock, soil, humus, etc. The gross appearance of these communities is similar throughout great areas of the world although species composition varies from region to region. The lichens may be restricted to specific substrates, but some species have such a wide amplitude that they may become established in several communities. The following discussions will outline the structure and dynamics of some typical communities.

CORTICOLOUS COMMUNITIES

Lichens are often the major components of epiphytic or bark-inhabiting communities in forested areas. Their chief competitors, at least at lower trunk levels, are bryophytes and, in tropical areas, epiphytic ferns, bromeliads and orchids. The most intensively studied areas are, of course, the temperate and subboreal forests where taxonomic knowledge has kept pace with phytosociological research. Little is known of lichen communities or taxonomy of the species in the vast tropical forests.

Classification of communities

The study of lichen communities has reached a high level of refinement, especially in Europe, and numerous phytosociological investigations have been published (see Barkman,[37] Klement,[159C] and Laundon[176A]). Lichenologists have in general followed the formalized systems that were first developed for classifying phanerogamic vegetation. In two major schools, one founded in Uppsala by Du Rietz and one in central Europe by Braun-Blanquet, all communities occurring on a given substrate (bark, soil, rocks, etc.) are grouped into classes. A further breakdown into orders is used if the substrate has a wide range of variation, such as living trunks, dead wood, etc. The main unit under order (and the highest unit in the Uppsala school) is the federation, designated by adding the suffix *-ion* to a lichen genus or species name, each federation in turn consisting of one or more unions denoted by the suffix *-etum*.

The actual field techniques are as follows. One visually surveys the lichen communities in a series of habitats and selects those that contain one or more constant or exclusive species for that union. A sampling frame, varying in size, dependent on the area of the substrate, is laid over the quadrat and cover of all species within the frame estimated on a scale of

1–10 (1 = one or two individuals and 10 = complete cover). The resulting data for a number of samples belonging to a union are assembled in tables (Table 7.1). Occasionally units different from those already described will be discovered, and these are proposed as new and given formal names.

Table 7.1 Fourteen samples of the *Caloplacetum heppianae* union (*Xanthorion* federation) in England.[176B]

	1	2	3	4	5	6	7	8	9	10	11	12	13	14
Caloplaca citrina	—	2	3	—	3	—	3	—	1	—	—	—	—	—
C. decipiens	—	—	—	—	—	—	4	4	—	—	—	—	8	—
C. heppiana	—	3	6	6	4	1	—	1	4	4	—	—	—	—
C. murorum	—	—	—	—	—	—	—	—	—	1	1	3	—	—
Candelariella aurella	—	4	1	3	1	3	—	3	—	3	2	2	—	3
C. medians	6	5	—	2	3	8	4	8	7	5	8	9	—	7
C. cf. vitellina	2	—	—	—	—	—	—	—	—	—	—	—	—	—
Lecanora dispersa	3	6	8	8	5	5	5	3	3	3	4	4	6	4
Physcia adscendens	—	—	—	—	—	—	3	—	—	—	—	—	—	—
P. caesia	5	4	—	—	—	5	4	—	5	4	—	—	—	—
P. orbicularis	—	—	—	—	—	—	—	—	2	3	—	—	—	—
Rinodina subexigua	—	—	—	—	—	—	—	—	—	—	—	1	—	2
Verrucaria nigrescens	—	—	—	—	4	—	—	—	—	—	—	—	—	—
Pyrenocarpic lichen	—	—	—	—	—	—	4	—	—	—	—	—	—	—
Alga	—	—	—	—	—	2	—	—	—	—	—	—	—	—
Total plant cover (%)	75	90	90	90	75	90	95	75	90	75	90	98	90	90

Laundon's[176A] discussion of the corticolous federations in England as summarized below will illustrate this system.

Cladonion coniocraeae

This federation of acid-loving lichens is confined to tree-bases, stumps, and acid peat. It is common in all temperate zones and especially so in eastern Britain. *Cladonia coniocraea* is the constant and exclusive species, by which the federation is recognized. Other species of common occurrence include *C. fimbriata*, *Lecidea granulosa*, and other crustose lichens.

Conizaeoidion

This federation is found only where the atmosphere is polluted as in highly urbanized or industrial areas. It is abundant over most of Britain and in many European cities. *Lecanora conizaeoides* is constant and exclusive in the federation.

Graphidion

This is a federation of shade-loving lichens found on smooth bark trees in the more temperate parts of Britain and Europe. Various species of *Graphis* are constant but other crustose lichens in the genera *Pertusaria*, *Lecanora*, and *Pyrenula* are often abundant and confined to this federation. Air pollution has apparently caused extinction of this community in many areas with replacement by *Conizaeoidion*.

Leprarion

This federation of shade-loving lichens also occurs on rough bark trees. The dominant constant species is *Lepraria*, a sterile crustose lichen. One union, the *Leprarietum*, is dominated by *L. aeruginosa* whereas the *Chaenothecetum* is dominated by *Chaenotheca melanophaea*, a leprose crust form in the Caliciaceae.

Lobarion

The characteristic habitat for this federation is moss-covered tree trunks in sheltered woods. Species of conspicuous foliose lichens *Lobaria*, *Nephroma*, *Pannaria*, and *Sticta*, are exclusive here. Only traces of the federation remain since it is sensitive to pollution.

Olivaceion

Communities on smooth bark, usually ash, dominated by *Lecidea olivacea* are classified in this federation. Various other crustose lichens occur in it: *Arthopyrenia*, *Lecanora*, and *Opegrapha*. This federation is favoured by well-lighted habitats and eventually replaced by *Graphidion* in shady situations.

Physodion

This very widespread federation grows on acid-barked trees in sunny habitats. Many unions have been described, based on the occurrence of showy foliose or fruticose lichens such as *Evernia prunastri*, *Parmelia caperata*, *P. sulcata*, *Menegazzia terebrata*, and *Usnea subfloridana*. *Hypogymnia physodes* is constant and exclusive.

Trichoterion

This federation is distinguished by the abundance of the exclusive species *Parmelia perlata*, often codominant with *P. caperata*. Species of *Ramalina* and *Usnea* may also occur. The typical habitat on exposed boles and branches is more severe than that of *Physodion*.

Xanthorion

This classic federation is characteristically found on neutral or basic barks and in habitats receiving a heavy supply of nitrates, such as farmyards, dirt roads, and bird perches. Species of the orange-coloured genus *Xanthoria* are constant and many other nitrophilous lichens occur here: *Physcia*, *Caloplaca*, *Buellia*, and some *Lecanora* species.

Establishment of communities

The most graphic example of invasion of trees by lichens is the colonization of the internodes of terminal branches. Many, though not all, trees have a rich twig flora, and the sequence of invasion by each species can be dated accurately from the annual terminal bud scars and growth rings. While there is some mention of this phenomenon in the older literature, the only comprehensive investigation of twig invasion was recently published by Degelius[79] for twigs of *Fraxinus excelsior* in Scandinavia.

According to Degelius' study, initial invasion of *Fraxinus* twigs begins on the third to fifth internodes. The first and second internodes have no visible trace of lichen growth. The first colonizers are not crustose species, as had previously been supposed, but foliose species, in particular *Xanthoria polycarpa* and *Physcia stellaris*. The first identifiable crustose species invade one to four years later, and fruticose species, as expected, develop last in the succession, usually not sooner than the tenth internode. *Fraxinus americana* in Minnesota has a similar orderly progression of colonization until more or less complete communities are established within 15 to 20 years (Table 7.1). A similar pattern of invasion has been followed through on terminal twigs of *Populus tremuloides*, where invasion is slower.[127] The twigs on this tree have longer internodes than *Fraxinus*; they are also more exposed because of a thinner canopy. Various species of *Quercus* in North America have a twig flora, but other trees, such as *Acer rubrum*, have none at all.

Vertical distribution of communities

Tree saplings are probably colonized by lichens initially through the same process as twig invasion. As the tree trunk and branches grow, the lichen communities form a more or less continuous cover over the bark. As the crown forms, however, the lower parts of the trunk become shaded and the environment along the trunk is radically altered, even more so as the crowns join as a continuous canopy.

The lichens become sorted out along the trunk in response to their different environmental needs, and a characteristic pattern of vertical distribution emerges. The floristic composition of communities at the tree base is quite unlike that in the crown. Xerophytic species rapidly disappear

Table 7.2 Frequency (%) of lichens invading internodes of *Fraxinus americana* in Minnesota. Sample size for internodes 1–10, 10 trees; for internodes 11 and 12, 8 trees; for internodes 13 and 14, 7 trees; and for internode 15, 4 trees.[127]

	Internode														
	1	2	3	4	5	6	7	8	9	10	11	12	13	14	15
Physcia stellaris	—	—	—	50	100	100	100	100	100	100	100	100	100	100	100
Xanthoria polycarpa	—	—	—	—	50	90	100	100	100	100	100	100	100	100	100
P. adscendens	—	—	—	—	—	10	40	50	80	90	100	100	100	100	100
P. orbicularis	—	—	—	—	—	10	20	40	70	90	100	100	100	100	100
P. ciliata	—	—	—	—	—	—	10	20	30	50	50	50	57	57	50
Lecidea melancheima	—	—	—	—	—	—	—	10	—	10	50	50	43	50	25
Caloplaca cerina	—	—	—	—	—	—	—	—	20	50	63	75	71	71	100
Lecanora subfusca	—	—	—	—	—	—	—	—	10	20	38	63	71	86	100
Candelariella vitellina	—	—	—	—	—	—	—	—	—	10	12	25	43	57	75
Rinodina pyrina	—	—	—	—	—	—	—	—	—	—	12	50	86	71	75
Candelaria fibrosa	—	—	—	—	—	—	—	—	—	—	25	12	14	—	25

and are replaced by shade-loving species of *Cladonia*, *Leptogium* and *Peltigera*. The canopy flora is rich in species of *Cetraria*, *Parmelia*, *Ramalina* and *Usnea*. Studies of this stratification (Fig. 7.1) in eastern North

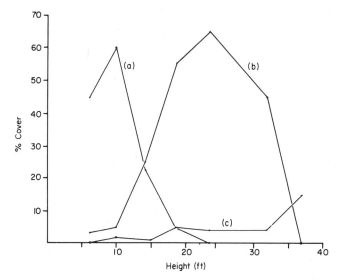

Fig. 7.1 Vertical distribution patterns of *Parmelia caperata* (a), *P. sulcata* (b), and *Ramalina farinacea* (c) on a 40 ft tall *Acer* at Dale Fort, Pembrokeshire, England. (After Kershaw[159])

America[110,126] and England[159] have clearly demonstrated this phenomenon. *P. subaurifera* and species of *Ramalina* are typical pioneer lichens, *P. sulcata* is the most frequent species in the mid to upper bole, and *P. caperata* dominates at lower levels.

The most useful aspect of vertical distribution studies is that a wide range of environment is compressed into the vertical height of a single tree, and the level of optimal development can be determined for each species by measuring coverage or frequency. Much remains to be learned about the factors that are correlated with vertical distribution of lichens. It is relatively easy to show gradients of light intensity and evaporation rate at various levels, but the mechanisms of colonization and the reason for the time differences in invasion are unknown. Brodo[51] shifted the positions of colonies of *Lecanora caesiorubella* from the 1.2 m level to base level on the same tree and reported continued growth. *Cladonia chlorophaea* shifted upward from base to 1.2 m soon decayed. Similar transplant experiments show promise of answering many questions on the amplitude of tolerance of lichens to environmental change.

Factors determining lichen distribution

Substrate specificity

While total forest environment has demonstrable effects on lichen distribution, it is also known that many lichens are tree specific. Conifers have a unique lichen flora rich in species of *Cetraria* and fruticose species of the Usneaceae, whereas deciduous hardwoods tend to have a preponderance of species of *Anaptychia, Parmelia, Physcia* and *Xanthoria.* In northern Wisconsin the differences in lichen frequency between *Acer* and the conifers *Tusga canadensis* and *Pinus strobus* are considerable (Table 7.3), and even the two main oaks in southern Wisconsin, *Quercus alba* with porous soft bark and *Q. velutina* with dense hard bark, have different frequencies of the common lichens.[111] Taking a small series of contiguous woodlots, Hale[127] concluded that about 60% of the variation in lichen communities could be ascribed to substrate factors and 40% to microclimate.

Table 7.3 Frequency per cent of major lichens on *Tsuga canadensis* (283 trees), *Acer saccharum* (125 trees) and *Pinus strobus* (330 trees) in northern Wisconsin.[67]

	T. canadensis	A. saccharum	P. strobus
Candelaria concolor	1	28	1
Cetraria ciliaris	3	1	31
Evernia mesomorpha	1	14	40
Graphis scripta	22	23	1
Lecanora subfusca	1	22	38
Parmelia aurulenta	1	50	0
P. caperata	15	40	76
P. galbina	1	40	4
P. physodes	8	2	37
P. rudecta	43	29	32
P. subaurifera	1	27	13
P. sulcata	5	43	36
Physcia orbicularis	1	56	1

Bark pH

There are a number of factors that can be correlated with differences in the bark substrate. These are summarized in great detail by Barkman.[37] One of the most intensively studied is pH or acidity of the bark. The range of variation between tree species is illustrated in Fig. 7.2 for some deciduous trees. Conifers generally have a lower pH, from 3–5. It should be noted that values even for the same tree species vary considerably and there is a broad overlap. Furthermore, pH of different levels on the tree trunk varies, the upper levels of bark having a higher pH than the lower. The presence of lichens on tree trunks modifies pH; Kershaw[159] found uncolonized areas to have a lower pH than colonized areas.

The most accurate and complete data on bark pH can still not enable us to decide how acidity affects lichen communities. A primary effect may be on the germination of fungal spores and growth of free-living algae, both of which have rather narrow limits of pH for optimal development when

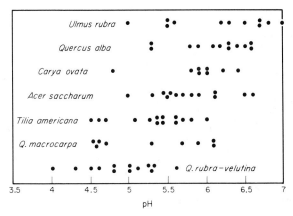

Fig. 7.2   pH of bark samples taken at 1.3 m on trees in southern Wisconsin. (From Hale[111])

grown in the laboratory. Kari[157] has discovered that the pH optimum of the algae does not correspond either to the pH of the lichen thallus or to that of the substrate. It is apparent that the pH effects on the alga would largely determine the success of lichenization. Algae normally require a higher pH than the fungi.

Water-holding capacity

Water-holding capacity is related to the porosity and texture of bark. In general soft-barked trees (*Ulmus, Fraxinus*) have higher water capacity and give up water more slowly than hard-barked trees (*Quercus, Carya*). Data on rate of evaporation for the common Wisconsin hardwoods (Fig. 7.3). show that the order of magnitude is not extremely large, certainly not enough to account for the differences in lichen communities. Measured bark hardness[67] parallels broadly the values for water-holding capacity, the hardest barks being least porous.

Other bark factors are difficult to assess and are relatively unstudied. For example, to what extent are nutrients leached from the bark during rainstorms and washed over lichens? Stemflow on trees in a temperate forest has been shown to be enriched, relative to throughflow, with potassium and calcium,[273] regardless of the tree species, and other properties of rainwater are probably modified.

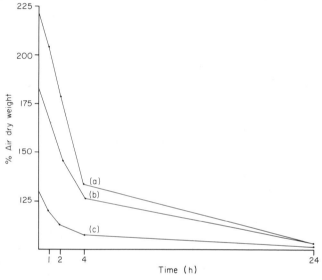

Fig. 7.3 Rate of evaporation from bark samples taken at 1.5 m from *Ulmus rubra* (a), *Tilia americana* (b), and *Acer saccharum* (c) in southern Wisconsin; atmospheric temperature 21 °C, relative humidity about 35%. (Adapted from Hale[111])

SAXICOLOUS COMMUNITIES

Classification of communities

Lichen communities on rock are classified in the same manner as corticolous ones described above on p. 89. Klement,[159C] for example, has described 12 federations in central Europe, including such widespread examples as *Rhizocarpion alpicolae*, *Umbilicarion cylindricae*, and *Parmelion saxatilis*. Laundon[176B] discusses in detail *Lecanorion dispersae*, a very common federation occurring on concrete and cement in urban areas. The characteristic species are crustose *Lecanora dispersa* and *Candelariella aurella*.

Establishment of communities

Xeric habitats

Lichens are conspicuous pioneers on rocks and frequently initiate stages of succession. As a substrate, a rock surface is far more stable than tree bark or soil and the lichen communities have a better chance of developing to maturity over long periods of time. The major environmental factor that alters the pattern of succession is shading by adjacent tree canopy and accumulation of leaf litter and soil debris that will modify or terminate the lichen communities. The principal cause of variation in the composition of saxicolous lichen communities is probably the type of rock substrate. Most lichens are rather strongly specific for certain kinds of rocks. The communities on basic rocks are dominated by species in the families Collemataceae, Physciaceae and Teloschistaceae; hard acidic rocks are frequented by the Parmeliaceae and Umbilicariaceae. Other factors not related to the chemical features of the rock can alter communities; the most notable is accumulation of nitrogen-rich excrement on rocks used by birds as perches.

While saxicolous communities have no patterning comparable to the vertical distribution of corticolous lichens, there are recognizable stages of succession. According to most studies, crustose lichens are the pioneers, followed by foliose and fruticose forms in an orderly succession. Beschel,[45] however, reported primary invasion by subfruticose *Alectorias* in the Arctic before a crustose lichen stage. Regardless of the first colonizers, the most important aspect in succession is what type of community can be considered 'climax' and how it is reached.

Several lichenologists have noticed that lichens do not follow an orderly progression to a 'climax' stage. There is sometimes a reversal in development, a non-directional succession, when foliose species are lost and replaced by crustose. This is, in theory at least, the reverse direction for succession. Moreau suggested the recycling of communities to explain this phenomenon.[1] Long-term studies of coverage of rock lichen communities in Connecticut[119] have provided more specific data on the dynamics of recycling. Foliose lichens, of which *Parmelia conspersa* is the primary component, increase steadily in coverage until the thalli reach a diameter of 9–10 cm and growth slows. At that time the older central parts of the thalli crumble and disintegrate. This degenerative process is usually hastened by an overgrowth of a leprous crustose lichen. Where the thallus has fallen away, fresh exposed rock surfaces are colonized by crustose lichens (*Lecanora, Rhizocarpon, Diploschistes*), which are often already present in low coverage but held in check by *Parmelia* or *Lepraria*. When observations are expressed graphically (Fig. 7.4), the cycling nature of this widespread rock community is manifest. Although individual thalli of *P.*

conspersa appear to live 30–40 years, the period of recycling seems shorter, 20–30 years.

Crustose colonies in the temperate forests of Connecticut are probably similarly altered in time. *Rinodina oreina*,[119] for example, has a much slower growth rate and, as one would expect, the recycling period is longer. Extrapolation from the average growth rate of 0.57 mm/year and the changes in coverage indicate a cycle of about 50 years. As in the case of *P. conspersa*, ageing and death of the central parts of the thalli accelerate the community changes. In the *Rinodina* community there is as yet no evidence of invasion by foliose species.

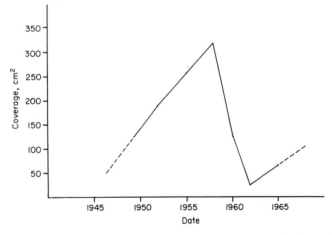

Fig. 7.4 Changes in coverage of *Parmelia conspersa* in a 567 cm² quadrat on gneiss in Connecticut, measured at 2–3-year intervals. Pecked lines indicate probable trends in coverage values. (Adapted from Hale[119])

Exposed rocks in glaciated areas of North America and Europe were uncovered some 10 000 years ago after the last glaciation. Since that time, recycling of communities must have occurred hundreds if not thousands of times, although there has been a continuous history of colonization. This phenomenon can explain why some exposed rocks are bare.

It should not be concluded that all saxicolous communities develop as rapidly as those in Connecticut. Frey's excellent photographs[97] show extremely little change for periods of up to 34 years for certain crustose–foliose rock communities. Furthermore, the persistence of small colonies of crustose lichens, especially species of *Lecanora* and *Rhizocarpon*, is noteworthy. As a generalization, however, it seems reasonable to assume that recycling will occur indefinitely until the micro-environment is so altered that mosses or other plants gain a foothold and crowd out the lichens.

Aquatic habitats

Lichens are not restricted to xeric habitats; they are of frequent occurrence in aquatic and semi-aquatic habitats, where they form distinct zones with relation to the water level. The communities are not necessarily confined to rocks; cast-up shells may also be invaded.

Santesson[231] conducted a large-scale study of lacustrine lichens in Sweden and found the lowest lichen zone, a blackish area submerged most of the year, to be composed of *Verrucaria* species. Above this was a brown zone of about 33 cm dominated by *Staurothele fissa* that was normally submerged about six months of each year. The next zone of 20 cm was occupied by light-brown *Lecanora lacustris*, which was submerged about three months of the year. The uppermost zone was dominated by *L. caesiocinerea* and was not immersed at all. Santesson correlated the lichen zones with water levels in the past and concluded that lichens are stable indicators of long-term changes in water level. In more recent work Ried[224A] found that differences in microclimate, temperature, light, and humidity between the zones are also highly correlated with community structure.

Similar zonation patterns of lichens along lake shores and rivers have been reported in North America[108,221] and Great Britain.[278] The submerged zone is almost always characterized by species of *Verrucaria*, *Arthopyrenia*, *Staurothele*, all Pyrenomycetes, or *Bacidia inundata*, a Discomycete. The partially submerged zone often contains *Dermatocarpon fluviatile*. Upper zones have *Lecanora* species and other crustose lichens.

Marine lichens form distinctively zoned communities along rocky seashores. Ecologists have divided the communities into as many as 21 separate zones and used different and sometimes conflicting terminology to describe them.[153] The simplest classification allows for three major zones, one below high-tide level and two above it. The intertidal zone is characterized by *Verrucaria maura* and marine species of *Arthopyrenia*, along with various marine algae and barnacles. Zones above this are not subject to normal tide action, but the lower part is affected by high waves and spray. This zone is recognized by the presence of black lichens, especially *Lichina* spp. and *Verrucaria maura*. Above this zone is a conspicuous orange belt dominated by *Caloplaca marina* and *Xanthoria parietina* as well as fruticose *Ramalina siliquosa*. Communities beyond this are out of reach of spray and intergrade with normal saxicolous communities.

SOIL COMMUNITIES

Klement[159C] has described six federations of soil-inhabiting lichen communities from central Europe. They are particularly well developed in alpine and arctic situations, prairies, and other habitats where trees are absent. *Cladonion* is probably the commonest federation, found in all vegetation zones.

Establishment of communities

Soil communities are highly vulnerable to environmental changes caused by growth or increase in density of grasses, sedges, shrubs or trees and accumulation of humus and leaf litter. If one traces the development of lichen communities from their initial invasion in open fields and the successional stages leading to forested stands, their transitory nature is at once evident. In the Piedmont region of south-eastern United States, lichens, primarily *Cladonia* species, invade abandoned fields and reach a peak of abundance between 10 and 40 years when the invading pines are 3–5 m high (Fig. 7.5).[225] The species appear to have some dependence on

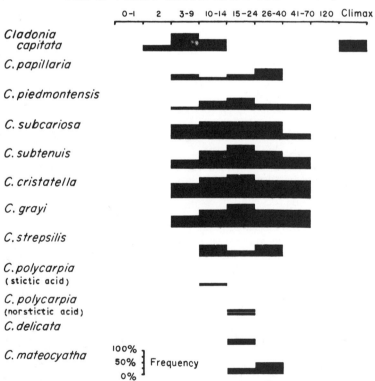

Fig. 7.5 Frequency of lichens in 2 × 6 ft quadrats laid in abandoned fields and reforested areas of known age in the Piedmont region of North Carolina. (From Robinson[225])

organic soil materials for establishment. After about 40 years, shading by the canopy and leaf litter have reduced the lichen cover and in stands 120 years old only one species of *Cladonia* was collected. Brodo[50] found a similar lichen succession on sandy areas in New York where after about 10 years of shading by the grass *Andropogon* a decline in lichen coverage was observed.

Lichens are an important agent in stabilizing soil. Several studies of coastal sand dunes in England[18] have delineated zones formed by as many as 13 communities. Strong correlation was found between the lichens and soil pH. Brodo[50] noted that *Cladonia boryi* follows the decay of *Ammophila breviligulata* in New York. In prairie and semi-desert regions, lichens, though inconspicuous, are frequently components of the ground vegetation. *Parmelia camtschadalis* grows loosely on soil as a pioneer lichen in large areas of Russia, North America, Australia and Africa. As grass cover increases, it is replaced by *Cladonia*.[185] Desert lichens, species of *Dermatocarpon*, *Heppia* and *Psora*, are important in consolidating soil.

The persistence of undisturbed alpine lichen soil communities is remarkable. Frey[97] photographed *Cladonia* carpets at a 20-year interval and could detect little change. To what extent this persistence can be attributed to antibiotic activity is a matter of conjecture. Inhibition of soil Phycomycetes by lichens is suspected,[129] and excretions from lichens might be effective in preventing or retarding seed germination or growth of encroaching higher plants.

Succession of lichens on soil is of special interest in subboreal areas where reindeer mosses and other lichens are important as fodder for domesticated animals. After normal foraging, a lichen pasture can regenerate in less than 20 years, but after total destruction by fire 80 years may be needed.[13] Conservation of this lichen resource in vast areas of North America and northern Europe, where lichens make up the primary soil communities, will undoubtedly be of great importance to ecologists in the future.

ANIMAL-INHABITING COMMUNITIES

It was long believed that lichens were incapable of growing on animals, for with their exceptionally slow growth they need a relatively stable substrate. Recently, however, the foliose lichen *Physcia picta* was collected on the shells of living giant Galápagos turtles, a suitably long-lived substrate.[133] However, as a more impressive example, entomologists have discovered large flightless weevils in the mossy forests of New Guinea that are thickly covered with lichens (Plate 16A), fungi, algae and liverworts.[105]

The lichens adhere closely to the pitted backs of the insects and are in turn infested with oribatid mites that appear to be lichenivorous and may help disperse the lichen propagules. The life span of these weevils has not been determined but, judging from expectable lichen growth rates, may exceed 10 years.

LICHENS AS EXTRA-TERRESTRIAL LIFE

In discussing the possible kinds of life that might inhabit other planets in the solar system, laymen frequently mention lichens as suitably primitive organisms. Salisbury[230] has concluded that lichens meet some but not all of the criteria that must be fulfilled by an inhabitant of Mars. They could withstand the hostile environment there, but could neither show the seasonal colour changes characteristic of Mars nor the ability to grow fast enough to cover areas so large as to be visible with a telescope. A well-informed lichen student could object to the idea that lichens are primitive; on the contrary, they are much more complex than other lower plants. Also, since lichens are dual organisms, we would have to presuppose the successful establishment and evolution of both fungi and algae on Mars, as well as subsequent lichenization, or the transplanting of lichens from earth. At the same time the most recent photographs and physical data from space-craft sent to Mars do not rule out the possibility of water in sufficient quantity to support micro-organisms. A final answer will come only after the Martian surface has been assayed by remote sensors or ultimately by man himself.

8

Chemistry of Lichens

INTRACELLULAR PRODUCTS

Intracellular products are bound in the cell walls and the protoplast. They are often water-soluble and can be extracted in boiling water. Since the lichen thallus is a composite structure, some of the products are synthesized by the fungus, some by the alga, and it is not always possible to decide what each component contributes. Most of the intracellular compounds isolated from lichens are non-specific and occur not only in free-living fungi and algae but also in higher green plants.

Carbohydrates

The hyphal cell walls of lichens are composed chiefly of lichenin, isolichenin, rarely pustulan, glucans or other as yet unidentified polysaccharides.[35] While many of these are unique to lichens, lichenin is reported from oat seeds where it is considered a 'reserve cellulose'.[197] It may have widespread and unsuspected occurrence in seeds. Lichenin is negative with iodine and yields glucose on complete hydrolysis. It is a linear polymer of β-D-glucose containing both 1,3- and 1,4-linkages in the proportion of 3:2 with a molecular weight of 20 000 to 40 000, placing it between starch and cellulose.[197] It is more readily digested by the enzyme lichenase than is cellulose and forms gels in water. Isolichenin, the rarer of the two major lichen starches, is distinguished by a positive iodine test and consists of D-glucose residues with α-1,3 and α-1,4 glucosidic linkages. Chitin is also a cell wall component of fungi but little information is available on its occurrence in lichens.

Low-molecular-weight carbohydrates are abundant in lichens and may constitute 3–5% of the dry weight of the thallus. Lindberg *et al.*[180] have analysed a number of Scandinavian species and found disaccharides (sucrose, α-trehalose, umbilicin) and polyols (erythritol, D-mannitol, volemitol, siphulitol). Mannitol is especially widespread in lichens. D-Arabitol appears to be a constant constituent of Discomycete lichens, whereas volemitol is confined to Pyrenomycetes, but too few lichen species have been investigated to decide if this is a valid correlation. It is fairly certain that mannitol is synthesized by the fungus, but either the fungus or the alga or both could be responsible for other sugar alcohols.

Carotenoids

Carotenoids are known to occur in lichens but very few species have been systematically analysed. Henriksson[137] found carotenoids in 16 species of the gelatinous lichen *Collema*, both in the thallus, where the colour is masked, and in the mycobiont cultures. Carotene is reported from *Roccella*.[202] As far as we know, water-soluble anthocyanins and flavonoids do not occur in lichens.

Free amino acids

Lichens are in general similar to other plants in their content of free amino acids. For example, leucine, isoleucine, phenylalanine, tryptophane, threonine, lysine, valine and methionine, all essential amino acids, have been extracted from thalli of *Peltigera canina*.[259] Various other amino acids reported are arginine, alanine, serine, glycine, glutamic acid, proline, sarcosine, α-aminobutyric acid and γ-aminobutyric acid, as well as the amide asparagine. A wider range of the commonly occurring amino acids and of other intermediates in protein synthesis would probably be shown by more extensive investigations on nitrogen metabolism.

Vitamins

A number of vitamins have been identified in lichen thalli, although their concentrations appear as a rule to be quite low. Several common arctic lichens have less than half the amount by dry weight of higher green plants.[107] These vitamins are reported: riboflavin, thiamine, niacin, ascorbic acid, pantothenic acid, nicotinic acid, folic and folinic acids, thymidine, biotin and other B_{12}-group vitamins.[107,246] Nearly all of them are presumed to be synthesized by the algae, for the fungi are usually poor sources of vitamins.

EXTRACELLULAR PRODUCTS

Extracellular products are deposited on the surface of the hyphae rather than within the cells. These products are usually insoluble in water and can only be extracted with organic solvents. With the exception of oxalates and carbonates that are widely distributed in the plant kingdom, the extracellular products of lichens constitute a large and unique group of compounds which have attracted the interest of many chemists.

Lichen substances

Although first discovered in the early part of the nineteenth century, the full extent and richness of the lichen substances were not appreciated until the work of the German chemists Hesse, beginning in 1861, and Zopf. Zopf[288] culminated descriptive investigations of the lichen substances in a monumental work *Die Flechtenstoffe*, published in 1907, in which he summarized the empirical formulae, properties and occurrence of over 150 compounds. While many of the compounds have since been found to be incorrectly determined or based on impurities or mixtures, this book is still a basic reference. The first laboratory synthesis of a lichen depside, lecanoric acid, was accomplished by Emil Fischer[93] in 1913, and since that time the molecular structures of over 80 other lichen substances have been established by Asahina in Japan,[33] Nolan in Ireland, Seshadri and Neelakantan[205] in India, Robertson in England, Lindberg and Wachtmeister[275] in Sweden, C. F. Culberson[65A] in America, Huneck[146A] in Germany, and others. The majority of the common substances are now more or less completely known, but many new compounds are still being discovered, especially among the depsides, depsidones, and triterpenoids.

The greatest number of lichen substances are weak phenolic acids synthesized only by lichenized fungi, but the following four pigments are also common to non-lichenized fungi: endocrocin and parietin, both anthraquinones, and polyporic and thelephoric acids, terphenylquinones.[245] A few other substances are structurally related to those in non-lichenized fungi. Picrolichenic acid, for example, has the same spirane nucleus as griseofulvin in *Penicillium* species.[276] Pyxiferin, a dibenzofurane pigment, is closely related to the fungal product oosporein. The only depsidone reported in non-lichenized fungi is a chlorinated β-orcinol derivative, nornidulin (XLII, Fig. 8.6).[245] Curiously, free monocyclic phenolic substances, including the presumptive precursors of depsides and depsidones such as orsellinic acid and sparassol, which are quite common in non-lichenized fungi, do not accumulate, since they are extremely rare in lichens.

Classification of lichen substances

The first major classification of lichen substances was proposed by Zopf[288] and later modified by Asahina and Shibata.[33] The substances were divided into two major series, aliphatic (fatty acids, polyols and triterpenoids) and aromatic (tetronic acid derivatives, depsides, depsidones, quinones, dibenzofuranes and diketopiperazine derivatives). Shibata[243] arranges the substances according to the biosynthetic pathways by which they appear to be synthesized. This newer system and the substances of known structure are given below. Newly discovered substances (since 1968) may be found in C. F. Culberson's catalogue,[65A] which also contains an exhaustive index of lichen species and their constituents. Other surveys of the substances by Shibata,[243] Neelakantan,[205] and Huneck[146A] may be consulted.

Outline of Shibata's Classification
   I. Shikimic acid origin
        1 Terphenylquinones
        2 Tetronic acid derivatives
  II. Mevalonic acid origin
        1 Triterpenoids
 III. Acetate-malonate origin
        1 Higher fatty acids
        2 Phenolcarboxylic acids
            (a) Orcinol derivatives: depsides, dibenzofuranes, depsones, dep-
                sidones, chromanones
            (b) β-orcinol derivatives: depsides, depsidones
            (c) Phloroglucinol derivatives
            (d) Quinones
  IV. Amino acid origin
        1 Diketopiperazine derivatives

I. SHIKIMIC ACID ORIGIN

1 Terphenylquinones

Thelephoric acid (I, Fig. 8.1).
Polyporic acid (II, Fig. 8.1).

These two deep-red or purple pigments are extremely rare in lichens. Thelephoric acid is deposited in the rhizines of *Lobaria retigera* and also occurs in a number of species of *Thelephora*, *Polyporus* and *Hydnum*, all non-lichenized Basidiomycetes. Polyporic acid is known from *Sticta coronata* and *Polyporus nidulans*.

2 Tetronic acid derivatives

Vulpinic acid (III, Fig. 8.1; R=H, M=COOCH$_3$)
Pinastric acid (III, Fig. 8.1; R=OCH$_3$, M=COOCH$_3$)
Pulvic anhydride (IV, Fig. 8.1)
Calycin (V, Fig. 8.1)

Leprarinic acid (VI, Fig. 8.1; M=H)
Leprarinic acid methyl ether (VI, Fig. 8.1; M=CH$_3$)
Rhizocarpic acid (III, Fig. 8.1; R=H, M=NHCHCOOCH$_3$CH$_2$C$_6$H$_5$)
Epanorin (III, Fig. 8.1; R=H; M=NHCHCOOCH$_3$CH$_2$CH(CH$_3$)$_2$).

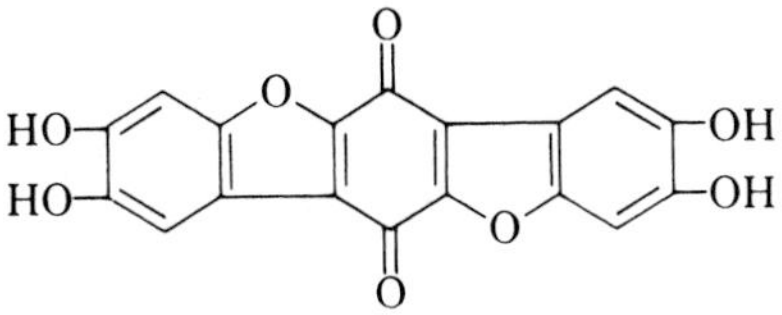

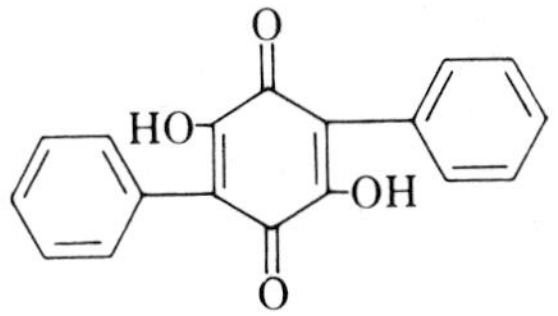

Fig. 8.1 Structural formulae of terphenylquinones and tetronic acid derivatives. See text for details.

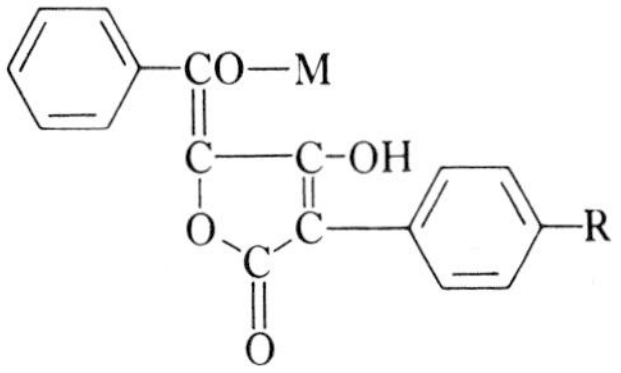

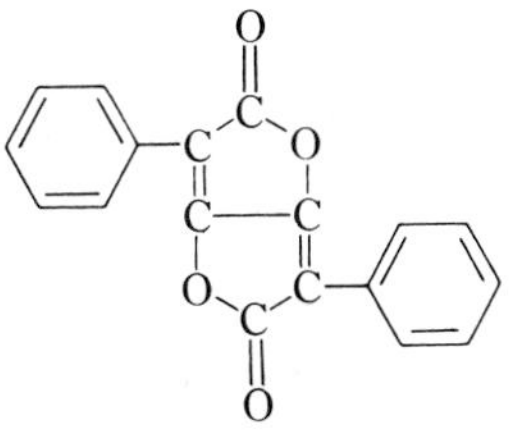

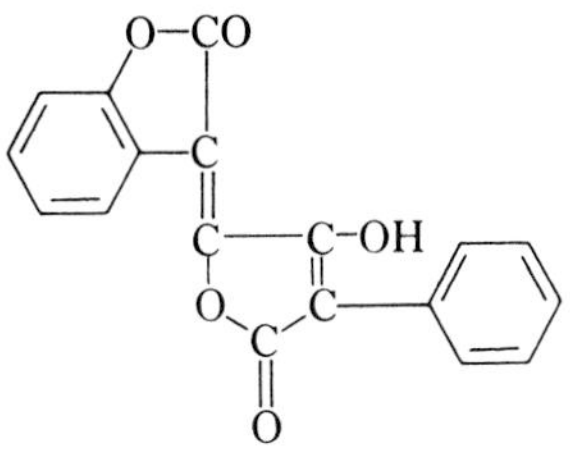

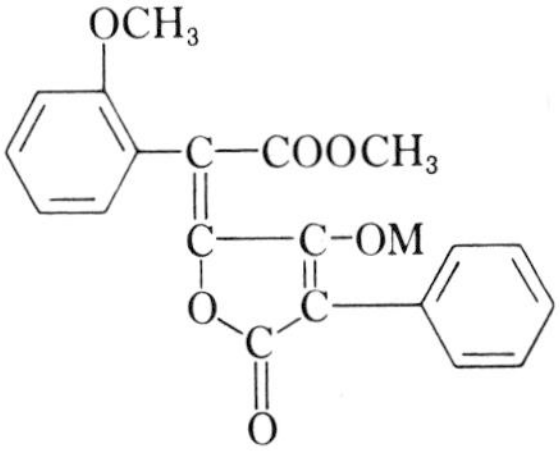

These substances were formerly called pulvic acid derivatives. They react with FeCl$_3$, a group indicator for phenolic rings, but are unreactive with KOH and Ca(OCl)$_2$. They are the yellow pigments in Caliciaceae, *Candelaria, Candelariella, Cetraria, Lepraria, Letharia, Pseudocyphellaria* and *Rhizocarpon*. Rhizocarpic acid and epanorin are unusual in being methyl esters of amino acids, L-phenylalanine and L-leucine respectively.

II. MEVALONIC ACID ORIGIN

1 Triterpenoids

Zeorin
Leucotylin
Ursolic acid
Taraxerene
Friedelin
epi-Friedelinol

These are colourless neutral pentacyclic compounds giving a positive reaction with acetic anhydride and concentrated sulphuric acid (Liebermann's test). They are relatively rare in lichens and occur in low concentrations. Ursolic acid is a component of *Cladonia* and has also been extracted from apple skins.

III. ACETATE-MALONATE ORIGIN

1 Higher fatty acids

(+)-Protolichesterinic acid, (−)-protolichesterinic acid, and (−)-allo-protolichesterinic acid (VII, Fig. 8.2; $R = CH_3(CH_2)_{12}$)
Nephrosterinic acid (VII, Fig. 8.2; $R = CH_3(CH_2)_{10}$)

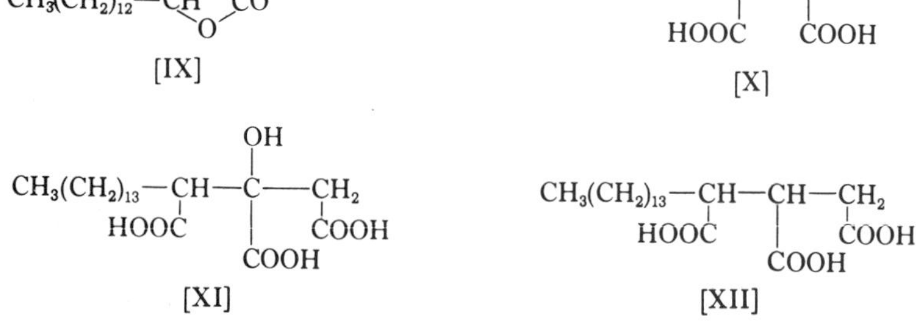

HOOC—CH—C=CH$_2$
R—CH   CO
      O
[VII]

HOOC—CH-HC—CH$_3$
R—CH   CO
      O
[VIII]

HOOC—C=C—CH$_3$
CH$_3$(CH$_2$)$_{12}$—CH   CO
            O
[IX]

CH$_3$(CH$_2$)$_{11}$—CH—CH—CH$_3$
         HOOC   COOH
[X]

                OH
CH$_3$(CH$_2$)$_{13}$—CH—C——CH$_2$
         HOOC | COOH
               COOH
[XI]

CH$_3$(CH$_2$)$_{13}$—CH—CH—CH$_2$
         HOOC | COOH
              COOH
[XII]

Fig. 8.2 Structural formulae of fatty acids. See text for details.

Nephromopsic acid (VIII, Fig. 8.2; $R = CH_3(CH_2)_{12}$)
Nephrosteranic acid (VIII, Fig. 8.2; $R = CH_3(CH_2)_{10}$)

(−)-Lichesterinic acid (IX, Fig. 8.2)
Roccellic acid (X, Fig. 8.2)
Caperatic acid (XI, Fig. 8.2)
Rangiformic acid (XII, Fig. 8.2)

Fatty acids of lichens are very closely related to those of fungi and widely distributed among all groups. None react with colour reagents. Aside from those listed above, a monocarboxylic acid, ventosic acid, and several tetrahydroxy-acids have been isolated from lichens but their occurrence is imperfectly known.

2 Phenolcarboxylic acids

(a) Orcinol derivatives

Depsides
Lecanoric acid (XIII, Fig. 8.3; $R=R'=CH_3$, $M=H$)
Erythrin (= the erythritol ester of lecanoric acid)
Evernic acid (XIII, Fig. 8.3; $R=R'=M=CH_3$)
Divaricatic acid (XIII, Fig. 8.3; $R=R'=C_3H_7$, $M=CH_3$)
Sphaerophorin (XIII, Fig. 8.3; $R=M=CH_3$, $R'=C_7H_{15}$)
Anziaic acid (XIII, Fig. 8.3; $R=R'=C_5H_{11}$, $M=H$)
Perlatolic acid (XIII, Fig. 8.2; $R=R'=C_5H_{11}$, $M=CH_3$)
Imbricaric acid (XIII, Fig. 8.3; $R=C_5H_{11}$, $R'=C_3H_7$, $M=CH_3$)
Diploschistesic acid (XIV, Fig. 8.3; $R=R'=OH$)
Obtusatic acid (XIV, Fig. 8.3; $R=OCH_3$, $R'=CH_3$)
Olivetoric acid (XV, Fig. 8.3; $R=R'=C_5H_{11}$, $M=M'=H$)
Microphyllinic acid (XV, Fig. 8.3; $R=C_5H_{11}$, $R'=CH_2COC_5H_{11}$, $M=CH_3$, $M'=H$)
Glomelliferic acid (XV, Fig. 8.3; $R=C_3H_7$, $R'=C_5H_{11}$, $M=CH_3$, $M'=H$)
Confluentic acid (XV, Fig. 8.3; $R=R'=C_5H_{11}$, $M=M'=CH_3$)
Sekikaic acid (XVI, Fig. 8.3; $R=R'=C_3H_7$, $M=M''=CH_3$, $M'=H$)
Ramalinolic acid (XVI, Fig. 8.3; $R=C_3H_7$, $R'=C_5H_{11}$, $M=CH_3$, $M'=M''=H$)
Boninic acid (XVI, Fig. 8.3; $R=C_3H_7$, $R'=C_5H_{11}$, $M=M'=M''=CH_3$)
Homosekikaic acid (XVI, Fig. 8.3; $R=C_3H_7$, $R'=C_5H_{11}$, $M=M''=CH_3$, $M'=H$)
Cryptochlorophaeic acid (XVI, Fig. 8.3; $R=R'=C_5H_{11}$, $M=M''=H$, $M'=CH_3$)
Merochlorophaeic acid (XVI, Fig. 8.3; $R=C_5H_7$, $R'=C_5H_{11}$, $M=M'=CH_3$, $M''=H$)
Methyl 3,5-dichlorolecanorate (XVII, Fig. 8.3)
Gyrophoric acid (XIX, Fig. 8.3; $M=M'=M''=H$)

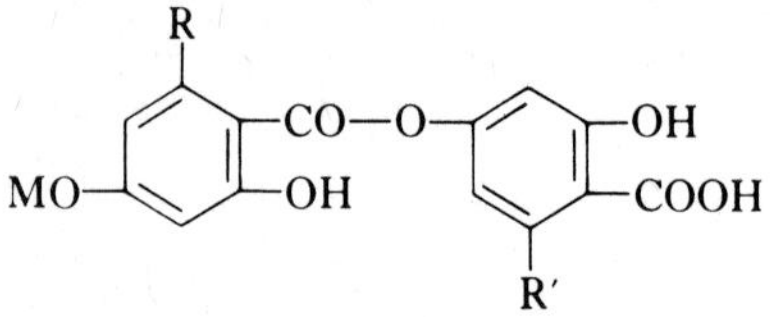

[XIII]

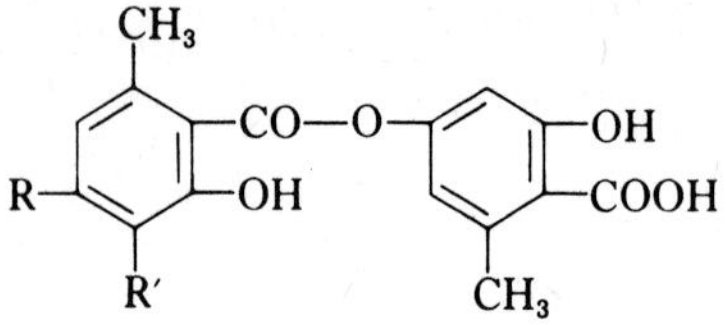

[XIV]

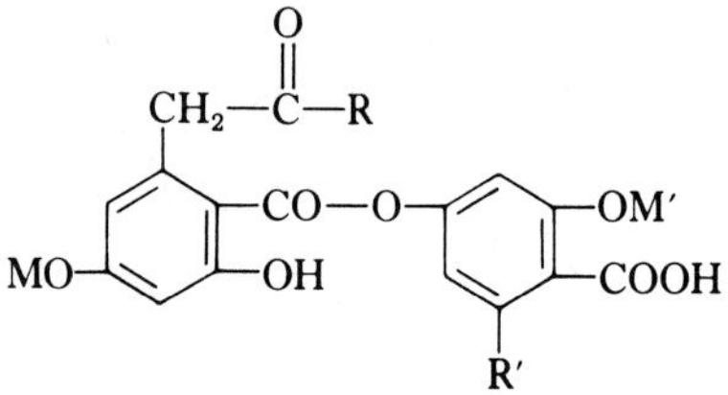

[XV]

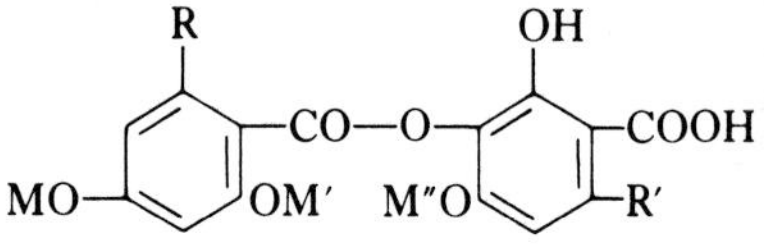

[XVI]

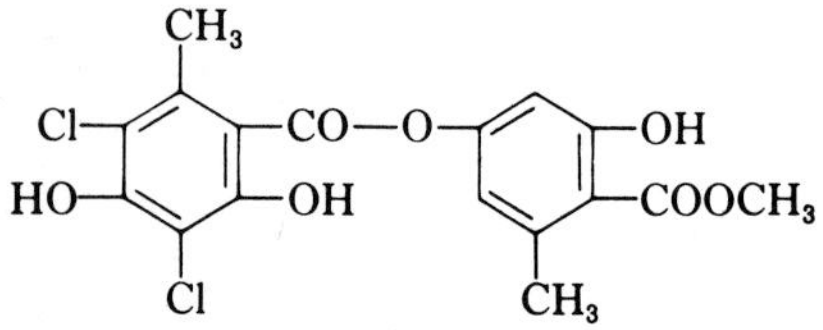

[XVII]

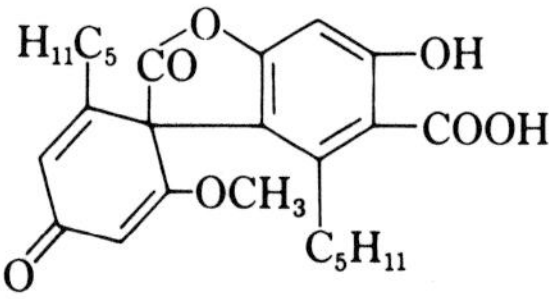

[XVIII]

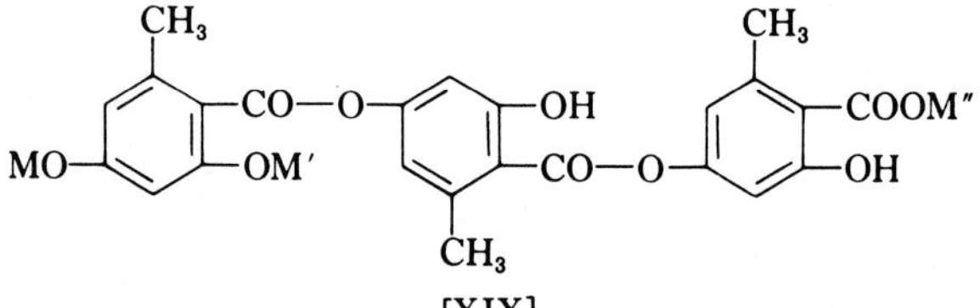

[XIX]

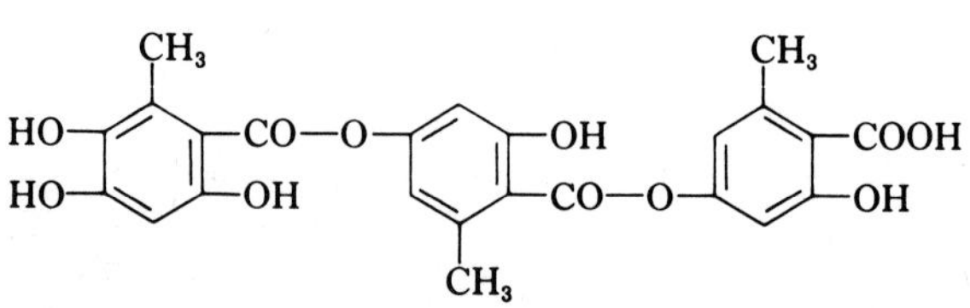

[XX]

Fig. 8.3 Structural formulae of depsides. See text for details. (XVIII) is the depsone picrolichenic acid.

Umbilicaric acid (XIX, Fig. 8.3; M=M″=H, M′=CH$_3$)
Tenuiorin (XIX, Fig. 8.3; M=M″=CH$_3$, M′=H)
Hiascic acid (XX, Fig. 8.3)

The depsides (including the β-orcinol derivatives listed below) make up the largest group of lichen substances. They are esters of phenolcarboxylic acids with a phenyl benzoate skeleton (Fig. 8.8), yielding two monocyclic halves on hydrolysis. They are colourless crystalline substances, reacting variously with FeCl$_3$, KOH, Ca(OCl)$_2$, and p-phenylenediamine. The depside nucleus is also known in digallic acid, a component of gallotannins which are widespread among higher plants.

Dibenzofuranes
Didymic acid (XXI, Fig. 8.4)
Strepsilin (XXII, Fig. 8.4)
Porphyrillic acid (XXIII, Fig. 8.4)
Pannaric acid (XXIV, Fig. 8.4)

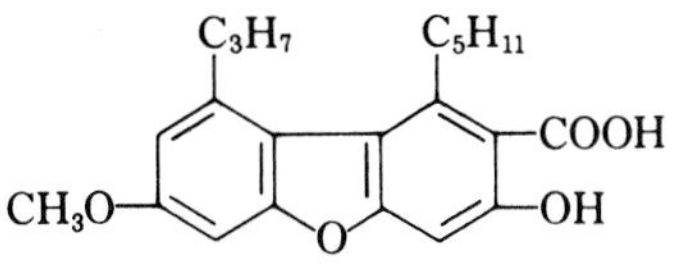

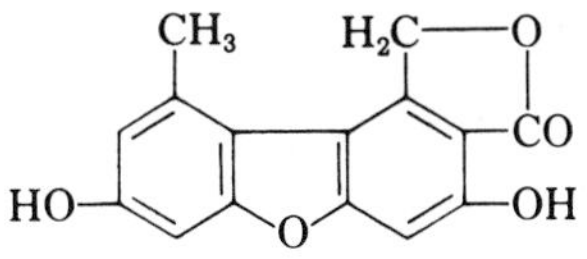

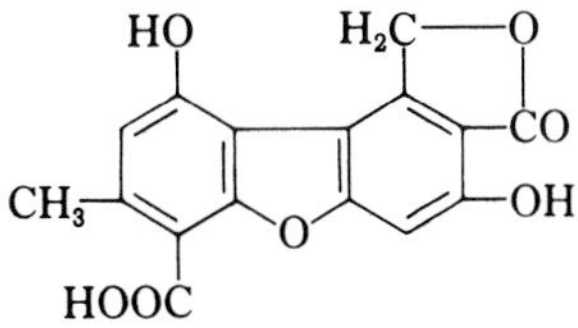

Fig. 8.4 Structural formulae of dibenzofurane derivatives. See text for details.

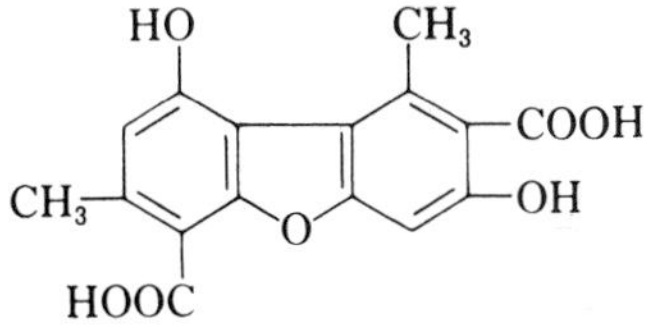

The dibenzofuranes are colourless rather rare compounds obtained almost exclusively from lichens. They may be considered derivatives of both diphenyl and diphenyl ether. Didymic acid and strepsilin are reported in several species of *Cladonia*, porphyrillic acid is in *Haematomma*, and pannaric acid is in *Crocynia*. These compounds are noteworthy for their antibiotic activity.

Depsones
Picrolichenic acid (XVIII, Fig. 8.3)

The characteristic structural feature of a depsone is the 3-spiro-2-one system. Picrolichenic acid, isolated from *Pertusaria amara*, is the only known representative of this class.

Depsidones
Physodic acid (XXV, Fig. 8.5; $R = CH_2COC_5H_{11}$, $R' = C_5H_{11}$, $M = H$)
α-Collatolic acid (XXV, Fig. 8.5; $R = R' = CH_2COC_5H_{11}$, $M = CH_3$)

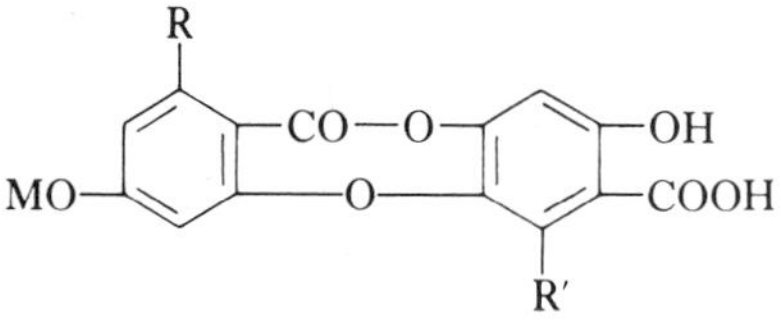

[XXV] 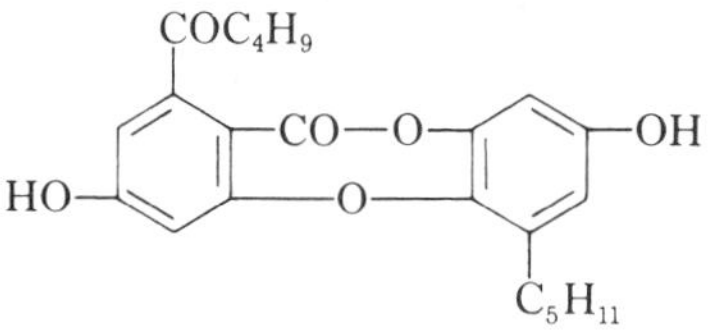 [XXVI] 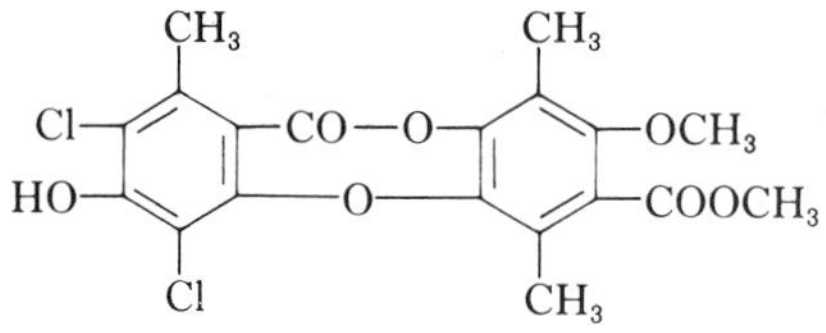 [XXVII] 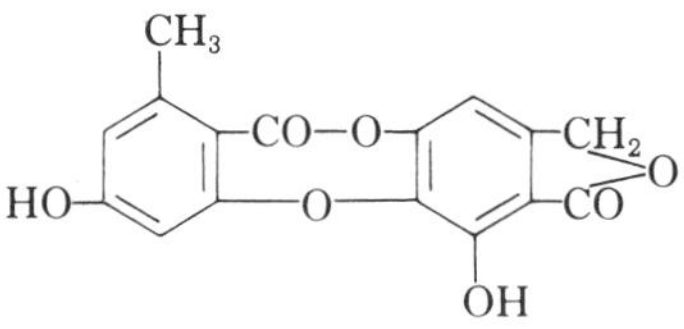 [XXVIII] 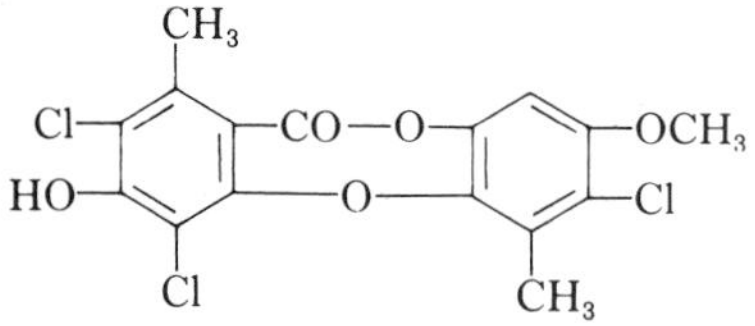 [XXIX] 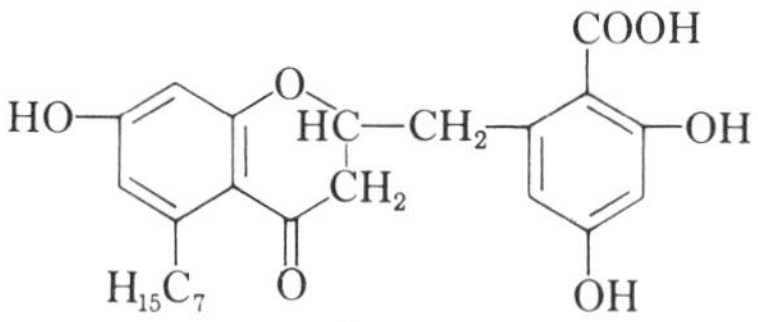 [XXX] 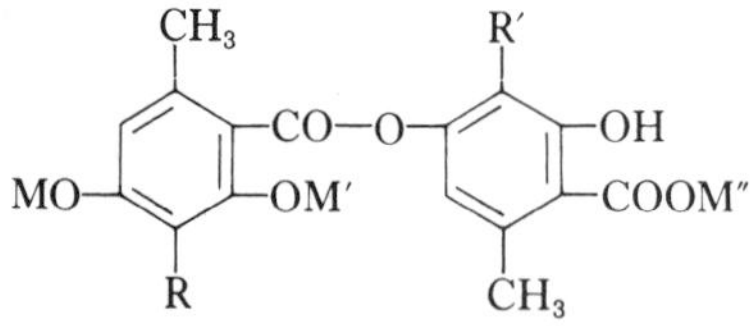 [XXXI] 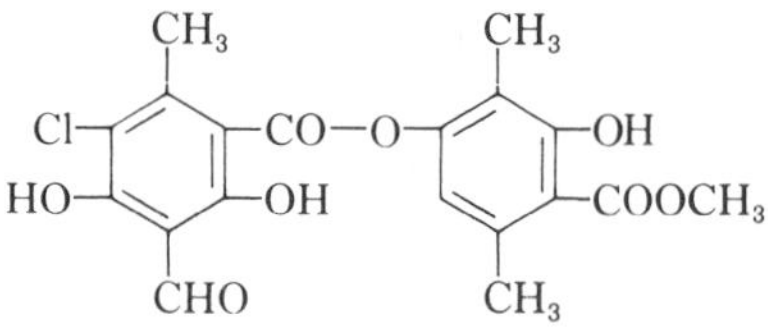 [XXXII]

Fig. 8.5 Structural formulae of orcinol derivatives. See text for details.

Alectoronic acid (XXV, Fig. 8.5; $R = R' = CH_2COC_5H_{11}$, $M = H$)
Lobaric acid (XXV, Fig. 8.5; $R = COC_4H_9$, $R' = C_5H_{11}$, $M = CH_3$)
Grayanic acid (XXV, Fig. 8.5; $R = CH_3$, $R' = C_7H_{15}$, $M = CH_3$)
Norlobaridon (XXVI, Fig. 8.5)
Gangaleoidin (XXVII, Fig. 8.5)
Variolaric acid (XXVIII, Fig. 8.5)
Diploicin (XXIX, Fig. 8.5)

Depsidones (including β-orcinol derivatives below) form the second largest group of lichen substances, similar to and probably biogenetically derived from depsides (see Fig. 8.8). The diphenyl oxygen linkage is very stable and on hydrolysis a depsidone yields only a hydroxy-diphenyl ether carboxylic acid derivative. Laboratory synthesis of these compounds is exceedingly difficult. Depsidones are colourless but can be identified with the usual colour test reagents.

Chromanones
Siphulin (XXX, Fig. 8.5)

This rare type of plant product has been isolated from *Siphula ceratites* in Norway.

(b) β-Orcinol derivatives

Depsides
Barbatic acid (XXXI, Fig. 8.5; $R = R' = M = CH_3$, $M' = M'' = H$)
Diffractaic acid (XXXI, Fig. 8.5; $R = R' = M = M' = CH_3$, $M'' = H$)
Atranorin (XXXI, Fig. 8.5; $R = CHO$, $R' = M'' = CH_3$, $M = M' = H$)
Baeomycic acid (XXXI, Fig. 8.5; $R = CHO$, $R' = M = CH_3$, $M' = M'' = H$)
Squamatic acid (XXXI, Fig. 8.5; $R = COOH$, $R' = M = CH_3$, $M' = M'' = H$)
Chloratranorin (XXXII, Fig. 8.5)
Thamnolic acid (XXXIII, Fig. 8.6; $R = CHO$)
Hypothamnolic acid (XXXIII, Fig. 8.6; $R = CH_3$)
Barbatolic acid (XXXIV, Fig. 8.6)

Depsidones
Salacinic acid (XXXV, Fig. 8.6; $R = CH_2OH$, $M = H$)
Norstictic acid (XXXV, Fig. 8.6; $R = CH_3$, $M = H$)
Stictic acid (XXXV, Fig. 8.6; $R = M = CH_3$)
Protocetraric acid (XXXVI, Fig. 8.6; $M = H$)
Fumarprotocetraric acid (XXXVI, Fig. 8.6; $M = COCH = CHCOOH$)
Physodalic acid (XXXVI, Fig. 8.6; $M = COCH_3$)

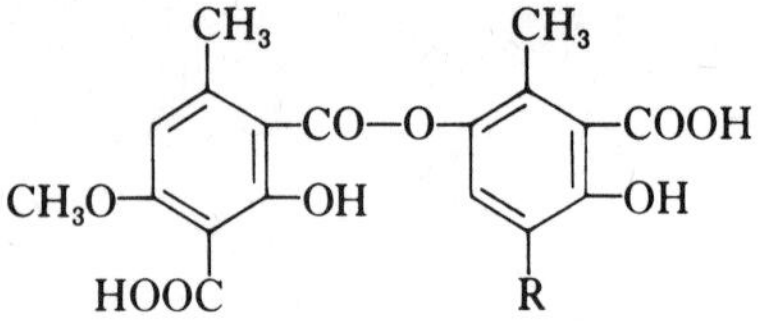

[XXXIII]

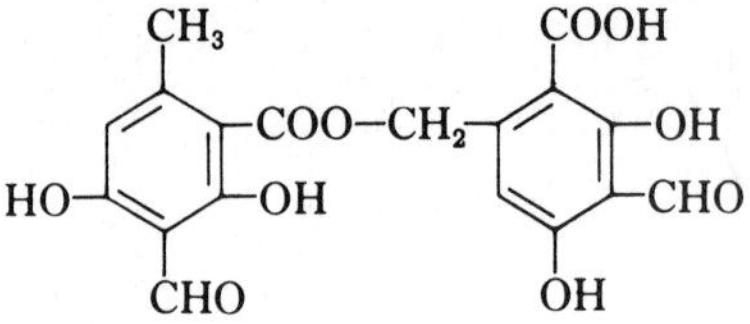

[XXXIV]

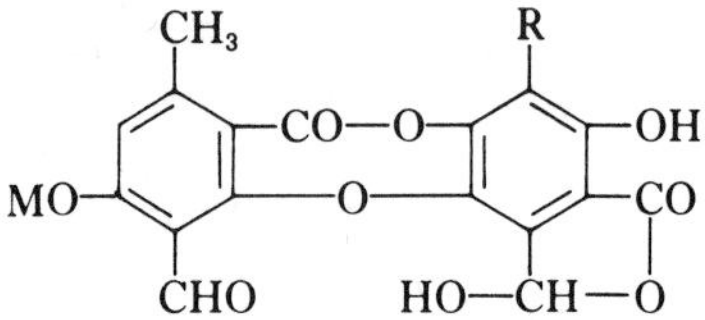

[XXXV]

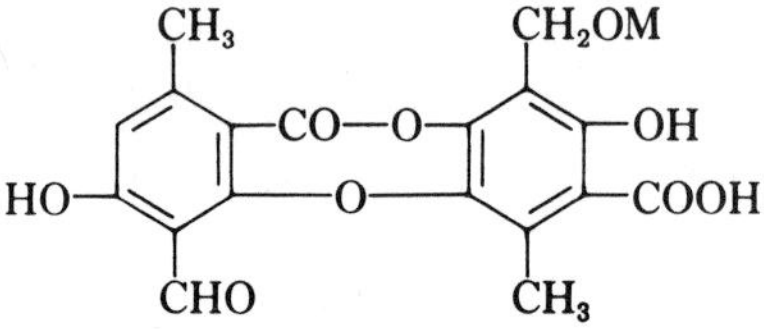

[XXXVI]

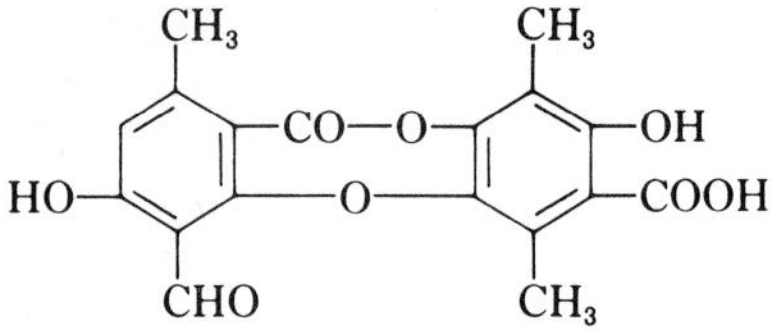

[XXXVII]

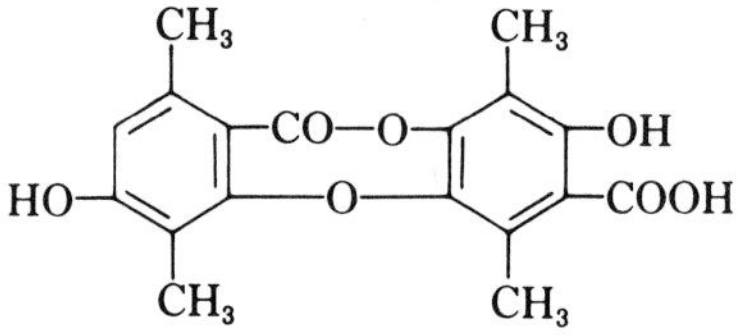

[XXXVIII]

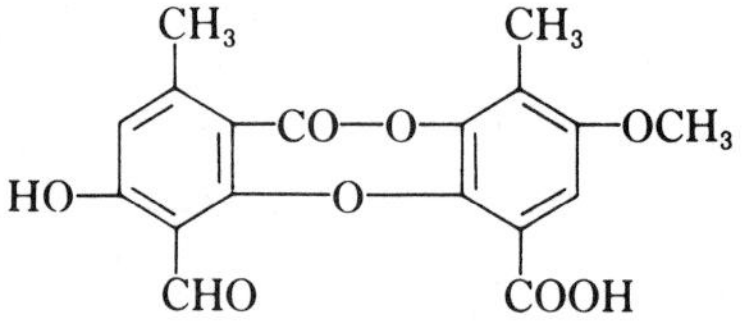

[XXXIX]

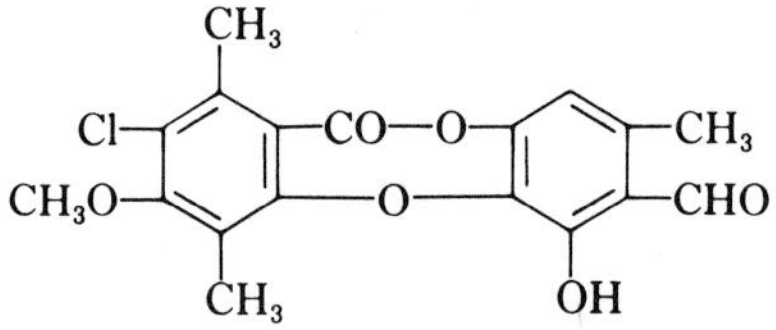

[XL]

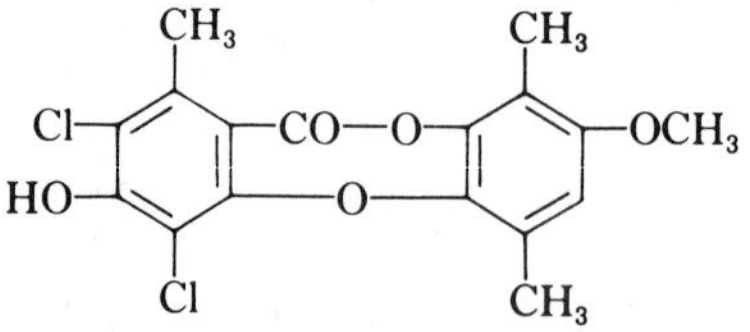

[XLI]

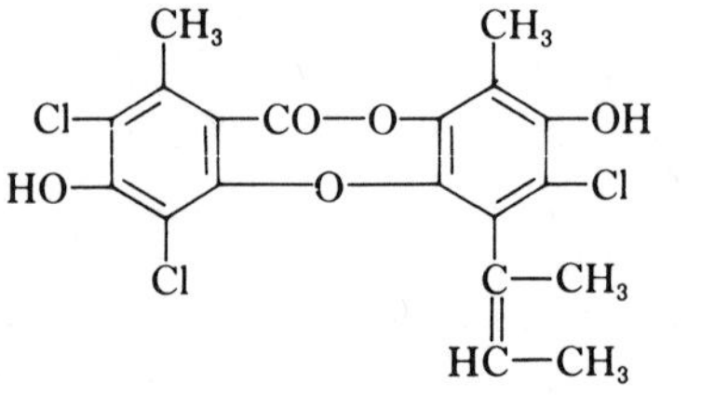

[XLII]

Fig. 8.6 Structural formula of β-orcinol depsides and depsidones. See text for details. (XLII) is nornidulin, produced by *Aspergillus*.

Virensic acid (XXXVII, Fig. 8.6)
Hypoprotocetraric acid (XXXVIII, Fig. 8.6)
Psoromic acid (XXXIX, Fig. 8.6)
Pannarin (XL, Fig. 8.6)
Vicanicin (XLI, Fig. 8.6)

Echinocarpic acid[163] and galbinic acid,[31] unidentified components of several *Parmelia* species, are classified as β-orcinol depsidones because of the P+, K+ colour reactions.

(c) Phloroglucinol derivatives
Usnic acid (XLIII, Fig. 8.7)
Lichexanthone (XLIV, Fig. 8.7)

Usnic acid, a pale yellow pigment, is sometimes placed with the dibenzo-furanes but does not fit well with any lichen substances. Its unusual chemical structure has intrigued many organic chemists and it has been the subject of a large body of literature. A very widespread substance, usnic acid is especially noteworthy for its commercial application as an antibiotic.

(d) Quinones
Dibenzoquinones
Pyxiferin (XLV, Fig. 8.7)

This deep-red pigment, recently isolated from the tropical lichen *Pyxine coccifera*, is the only representative of dibenzoquinones in lichens.[205] This type of quinone, and also monobenzoquinones, are much more common in non-lichenized fungi.

Anthraquinones
Parietin (physcion) (XLVI, Fig. 8.7; R=CH₃)
Fallacinol (teloschistin) (XLVI, Fig. 8.7; R=CH₂OH)
Fallacinal (XLVI, Fig. 8.7; R=CHO)
Parietinic acid (XLVI, Fig. 8.7; R=COOH)
Endocrocin (XLVII, Fig. 8.7)
Rhodocladonic acid (XLVIII, Fig. 8.7)
Solorinic acid (XLIX, Fig. 8.7)

Anthraquinones are rather closely interrelated orange or red pigments that react purple with KOH. They figure prominently as pigments in fungi, higher plants and insects. Among the lichens, a number of suspected anthraquinones remain to be analysed. For example, *Herpothallon sanguineum*, a common crustose lichen in tropical America, has a deep purple-red pigment called chiodectonic acid. Rhodophyscin is reported in *Physcia* and *Parmelia* but its structure is unknown.[124]

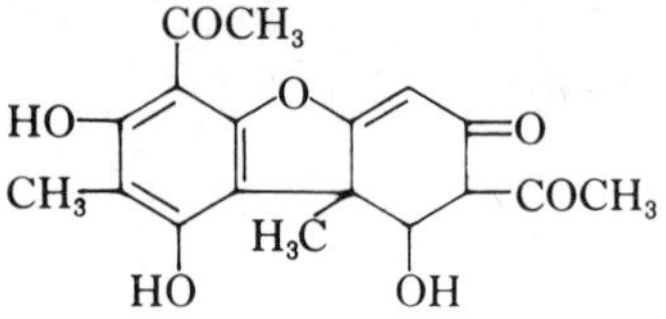

[XLIII]

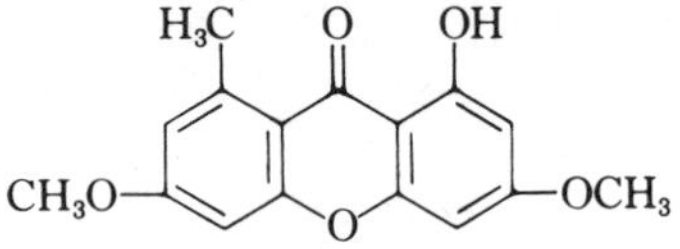

[XLIV]

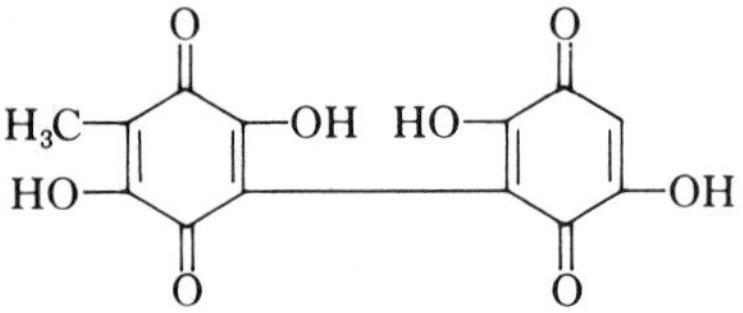

[XLV]

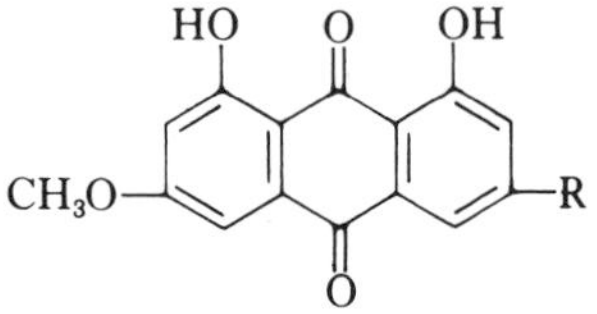

[XLVI]

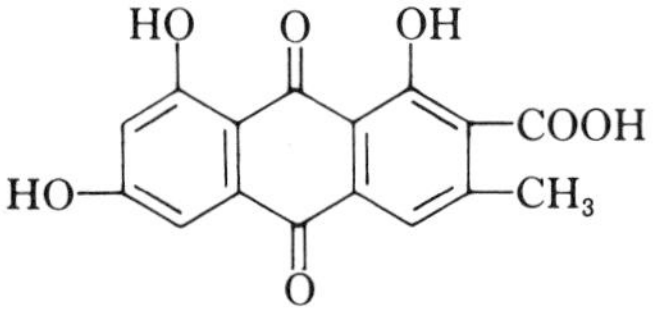

[XLVII]

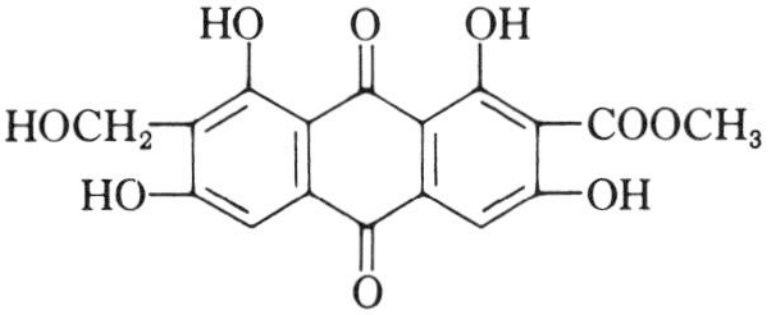

[XLVIII]

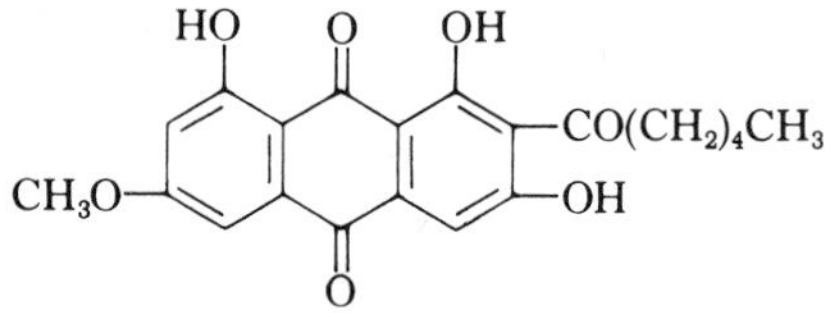

[XLIX]

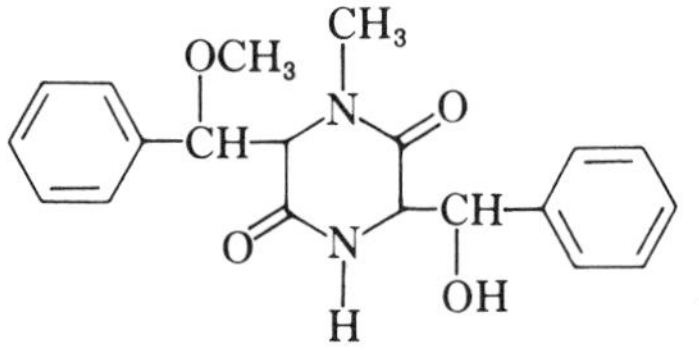

[L]

Fig. 8.7 Structural formulae of quinones, phloroglucinol derivatives, and diketopiperazine derivatives. See text for details.

IV. AMINO ACID ORIGIN

1 Diketopiperazine derivatives

Picroroccellin (L, Fig. 8.7)

This curious derivative has been found only in *Roccella fuciformis* and is a rare type of compound in nature. It is apparently derived from two amino acids, β-phenylserine and *O,N*-dimethyl-β-phenylserine, joined in a diketopiperazine ring structure.

Biosynthetic pathways

Mosbach[201] has recently completed the first biosynthesis of the mono-cyclic phenolic precursors of depsides in the laboratory. Thalli of *Umbili-caria pustulata* were placed in a solution of diethyl malonate $1-^{14}C$ in shake culture for several hours. In this brief time the lichen assimilated some of the radioactive metabolite and incorporated it into gyrophoric acid, a normal constituent of this lichen. The acid was extracted, hydrolysed and analysed. Mosbach postulates that orsellinic acid, the basic building block for gyrophoric acid, is synthesized from one molecule of acetyl coenzyme A and three molecules of malonyl coenzyme A, following a typical acetate-polymalonate pathway that is characteristic of other micro-organisms. A similar pathway, but with reduction instead of cyclization, could explain the synthesis of fatty acids.

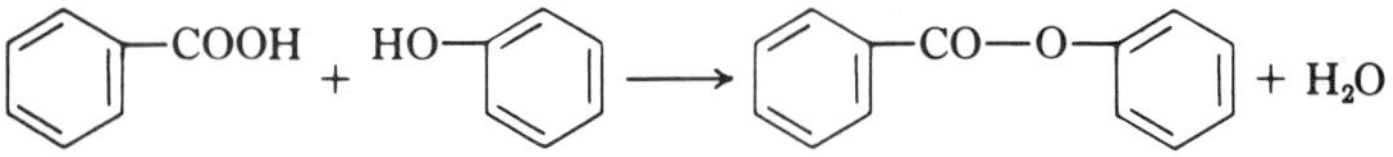

[I]

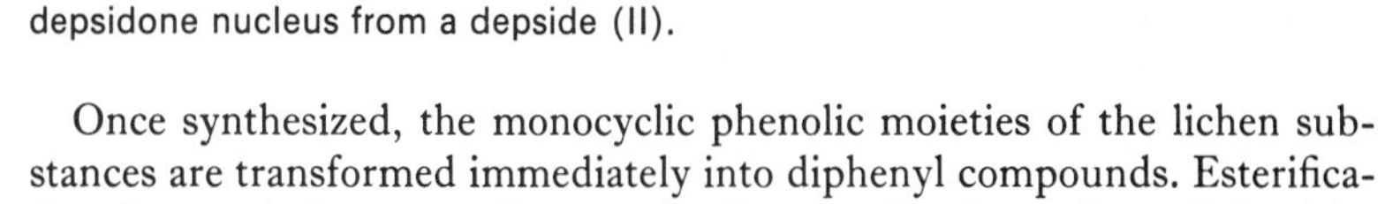

[II]

Fig. 8.8 Formation of the depside nucleus by esterification of a phenylcarboxylic acid and a phenol (I) and the mechanism of oxidative cyclization to derive a depsidone nucleus from a depside (II).

Once synthesized, the monocyclic phenolic moieties of the lichen sub-stances are transformed immediately into diphenyl compounds. Esterifica-tion is an obvious coupling mechanism in the formation of depsides (Fig. 8.8), but it is more difficult to account for the ether linkage in depsi-dones. From evidence based on the first successful synthesis of a depsidone, diploicin, Brown *et al.*[55] are of the opinion that the diphenyl ether linkage arises by oxidative transformation (Fig. 8.8). If so, it is reasonable to assume that depsidones are derivable *in vivo* from depsides. Differences between the depsides and depsidones probably arise from nuclear hydroxy-lation (diploschistesic acid from lecanoric acid), *O*-methylation (β-collatolic acid from alectoronic acid), *C*-methylation (obtusatic acid from evernic acid), and similar biosynthetic mechanisms. Polyporic acid or its derivatives seem to be the precursors of both pulvic anhydride[189] and calycin.[15] C. F. Culberson[65A] and Huneck[146A] have both summarized recent work and theories on the biosynthesis of lichen substances.

Concentration in the thallus

Concentrations of the lichen substances in the thalli are often quite high in comparison with non-lichenized fungi or higher plants, although there is considerable variation. The values shown below are percentages of thallus dry weight for some of the commoner substances in various lichens:

Atranorin	$1.2,^{28}$ 2.8^{115}
Cryptochlorophaeic acid	0.16^{244}
Divaricatic acid	$1.4,^{112}$ 2.5^{146}
Fumarprotocetraric acid	0.55^{28}
Gyrophoric acid	$10,^{148}$ $1.0-3.59^{194}$
Lecanoric acid	4.3^{112}
Merochlorophaeic acid	0.28^{244}
Norstictic acid	6.0^{149}
Parietin	1.4^{112}
Roccellic acid	4.2^{147}
Salacinic acid	6.0^{127}
Usnic acid	$0.2-4.0^{234}$
(+)-Usnic acid	$0.5-1.8^{257}$
(−)-Usnic acid	$0.22-8.0^{257}$

The substances are deposited in the medulla or in the cortex, rarely in both layers, with a high degree of specificity. Typical cortical substances, most anthraquinones, tetronic acid derivatives and usnic acid, are pigmented, but atranorin and lichexanthone are colourless. By contrast, most depsides and depsidones are deposited only in the medulla. The substances are infrequently concentrated in the ascocarps or in the soredia but absent or less concentrated in the main thallus.

Role of lichen substances

The uniqueness of lichen substances has stimulated much speculation on their physiological role and evolutionary significance. In the case of non-lichenized fungi, insoluble metabolic shunt products often serve as reserve food and are utilized when exogenous sources of carbon are low. It is not at all certain, however, that lichen substances, once deposited on the hyphae, are available for further metabolism. Gyrophoric acid in *Umbilicaria pustulata* is maintained at the same concentration as the thallus deteriorates and dies,[194] an indication that the acids are not resorbed by the hyphae. According to Culberson and Culberson,[66] the percentage dry weight content of gyrophoric acid in *U. papulosa* is constant for all age classes of thalli investigated. This finding suggests that the deposition of acids on the hyphae is carefully regulated by the lichen and that indefinite accumulation does not occur with maturation.

Lichen acids and pigments increase the opacity of the upper cortex.[88] This reduction of light reaching the algal layer may perhaps benefit *Trebouxia*, which seems to tolerate lower light intensities than non-lichenized algae. *Trebouxia* would have a better chance of survival in a highly insolated xeric habitat if protected by opaque hyphae. An accessory physiological function has been suggested by Rao and LeBlanc,[219] who discovered that atranorin, a cortical substance, acts as a fluorphor emitting fluorescence at a peak of 425 nm, coinciding with the absorption spectrum of chlorophyll. Thus atranorin or other cortical substances may act as light absorbers and increase the ability of the phycobiont to use light of shorter wavelengths.

Since medullary hyphae heavily encrusted with acid crystals are not easily wetted, the acids perhaps regulate or prevent excessive water uptake, although lichens, normally so limited by availability of water, would hardly seem dependent on such a mechanism. The bitter taste of some of the acids, especially fumarprotocetraric, may make lichens unpalatable for some animals, but reindeer are not deterred from consuming large quantities of bitter reindeer mosses. A more realistic assessment of a protective role for lichen substances might be based on their antibiotic properties. Lichens are seldom attacked by bacterial or fungal pathogens in nature, even though they may have life spans of 50 to 100 years. The antibiotic activity has been correlated with the presence of specific lichen acids.[143] In the herbarium, where higher plants and fungi are so susceptible to insect infestation, most lichens can be preserved with only minimal precautions for fumigation.

Identification of lichen substances

Colour tests

Colour tests were discovered and systematically used by Nylander as long ago as 1866.[208] He employed two reagents, bleaching powder (Ca(OCl)$_2$) and potassium hydroxide (KOH). By convention calcium hypochlorite is abbreviated as C and potassium hydroxide as K. Nylander used the reagents separately as well as in a combination of K followed immediately by C, the so-called KC test. The only addition to this useful trio of tests, *p*-phenylenediamine (P), was made by Asahina. The P test is highly specific for most substances that react K+ yellow or red, and has largely supplanted the simple but sometimes indecisive K test. Both the cortex and medulla are tested; positive reactions are indicated as K+, C+, KC+ or P+, in that order, and if negative, as K−, C−, KC− or P−. A positive reaction with K gives a yellow coloration or yellow turning slowly to red; with C, red, rose or orange red; with KC, red or rose; and with P, brick red, orange or yellow.

It is essential that the solutions be prepared with special care to ensure accuracy. *p*-Phenylenediamine should be made up fresh as a 5% alcoholic solution; it deteriorates within a day. A stable P solution (Steiner's solution) can be prepared by combining 100 cm³ water, 10 g sodium sulphite, 1 g *p*-phenylenediamine, and 40 drops of any liquid detergent. This mixture is slower to react than the alcoholic solution. Calcium hypochlorite must be prepared from active chemical stock as a saturated aqueous solution; it decomposes in a day or two. Potassium hydroxide can

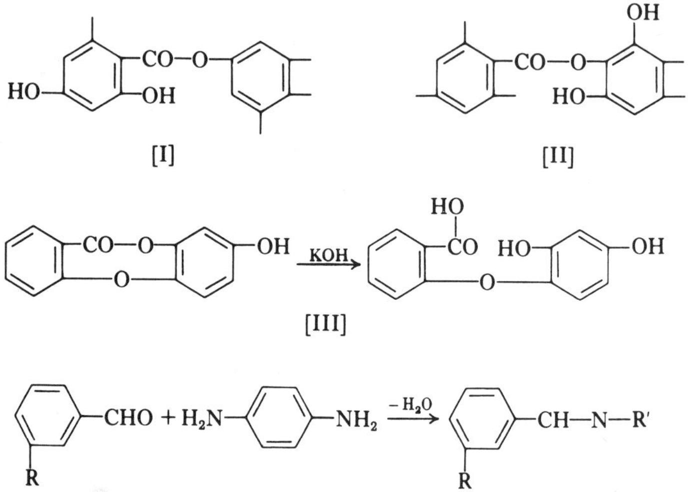

Fig. 8.9 Basic structure of 2,4 *m*-dihydroxy (I) and 2',6' *m*-dihydroxy (II) depsides that react with $Ca(OCl)_2$. Cleavage of the orcinol depsidone ester linkage by hydrolysis (III) with formation of 2',4' *m*-dihydroxy-derivative. Formation of Schiff bases by reacting an aldehyde radical on a depside or depsidone nucleus with *p*-phenylenediamine (IV); R = depside or depsidone moieties, $R' = C_6H_5NH_2$.

be made up simply as a 10% aqueous solution that is stable. If the test reactions are checked with lichens of known chemical composition, there will be no doubt as to the reliability of the solutions.

Colour reactions of the lichen substances are predictable from the molecular structures. The C test is positive in acids that have a *m*-hydroxy-configuration (I and II, Fig. 8.9). Treatment with a mild oxidizer may induce the formation of pigmented *o*-quinones (as in lecanoric, anziaic, gyrophoric, olivetoric and hiascic acids) or *p*-quinones (as in ramalinolic and cryptochlorophaeic acids) (Fig. 8.9), but no such reactions have been proved experimentally. The colours of the quinones, orange to deep red, are probably modified by other radicals present on the phenyl rings. The positive C test of methyl 3,5-dichlorolecanorate and several other substances defies explanation.

The KC+ red or orange-red reactions of orcinol depsidones (physodic, alectoronic, lobaric acids and norlobaridon) probably have similar quinone formation after the ester linkage is cleaved by hydrolysis in alkali (III, Fig. 8.9). Quinones are apparently blocked by methyl or other side chains positioned between the hydroxyls in β-orcinol-derived depsidones (salacinic acid, stictic acid, etc.), all of which are negative with C.

A positive reaction with p-phenylenediamine is specific for aldehyde (CHO) radicals on both depsides and depsidones. The mechanism appears to be a condensation reaction resulting in yellow, orange or red Schiff bases (IV, Fig. 8.9), but evidently with a considerable degree of conjugation between the phenyl rings. Many of these same acids react with KOH but the mechanism entails the formation of unidentified complex ions.

Fluorescence analysis

Examination of lichen thalli under long-wave ultraviolet light is another test for spotting groups of acids with similar configurations. The following depsides and depsidones show a bright white to bluish- or greenish-white fluorescence, in order of intensity[113]: alectoronic acid, squamatic acid, sphaerophorin, lobaric acid, divaricatic acid, evernic acid, psoromic acid, perlatolic acid and barbatic acid. Other reactive substances are rhizocarpic acid and lichexanthone which fluoresce brilliant orange.

Ultraviolet absorption spectra

A final method for identifying related groups of lichen substances is through their ultraviolet absorption patterns. This advanced technique requires that the substances be highly purified and has therefore been utilized to only a limited extent by taxonomists. Chemists use it to identify new or unknown substances. Patterns of absorbence for some of the major groups are as follows[116]:

Fatty acids: Lactonic acids—strong absorbence in short wavelengths; open chain acids—no absorbence.

Tetronic acid derivatives: Uniformly strong absorbence from 220 nm into red wavelengths.

Depsides and depsidones: Strong absorbence in two regions, 300–310 nm and 240–260 nm, giving more or less bimodal curves; peaks dampened as side chains and oxidation levels increase.

Microchemical tests

CRYSTALLIZATION   The colour tests with K, C and P have been employed by lichenologists for many years and have amply proved their usefulness in practice. However, these tests are non-specific in the sense that they identify only certain structural configurations. Different acids that share the same colour-reacting radicals will give the same colour test and cannot be distinguished. For example, all free *m*-hydroxy depsides (lecanoric acid, olivetoric acid, etc.) will react C+ red, and all of the common orcinol depsidones are KC+ red. Since many acids do not react with any colour reagents, there is no way to identify them with colour tests.

By 1900 some crude microchemical tests had been devised for lichen substances, but it remained for the Japanese chemist Asahina[21] in 1936 to simplify and standardize tests that can be performed on a microscope slide with a high degree of accuracy and speed. The substances are identified by characteristic crystal formation in various reagents. These tests have brought identification of the lichen substances within the easy reach of taxonomists and physiologists. To some extent, of course, microchemical tests compromise reliability for the sake of simplicity of method and materials. Interesting or unusual findings first reported from microchemical studies should be verified by macro-extraction when sufficient material becomes available.

The technical steps in making a microchemical test are relatively simple.[120] A fragment of the lichen thallus about 1 cm^2 in area is crumpled and placed on a microscope slide (Fig. 8.10). Drops of acetone or other organic solvents as required are pipetted over the fragments and quickly dissolve out any lichen substances present. A ring or residue of crystals form around the base of the fragments on drying. The thallus fragments may also be extracted in small vials and the concentrated solution pipetted directly on a microscope slide. After the expended fragments and debris are brushed away, a coverslip is placed directly over the residue and a crystallizing reagent added. After moderate heating, the lichen substances recrystallize in distinctive shapes and colours that can be identified by comparison with photographs (Plates 14, 15).

Crystallizing reagents first introduced by Asahina have been standardized as follows, using a volume basis:

G.E.: Glycerine-acetic acid, 1:3.
G.A.W.: Glycerine-alcohol-water, 1:1:1.
G.A.*o*-T.: Glycerine-alcohol-*o*-toluidine, 2:2:1.
G.A.An.: Glycerine-alcohol-aniline, 2:2:1.
G.A.Q.: Glycerine-alcohol-quinoline, 2:2:1.

There is unfortunately no adequate summary of the microchemical methods. Specific details must be sought in Asahina's scattered publications[21] and in briefer summaries by Evans,[89] Hale[120] and Krog.[161] The following key and brief descriptions of the crystals of the commoner substances may prove helpful for preliminary investigations. Representative lichen species that are known to contain the substances are also listed.

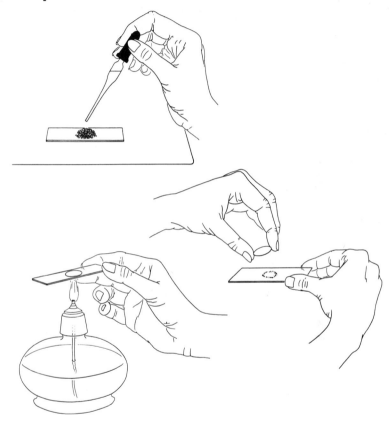

Fig. 8.10 Steps in making a microchemical test : extraction of thallus fragments, adding crystallizing reagents with coverglass, and heating gently over a spirit lamp. (Drawings by N. Halliday)

KEY TO LICHEN SUBSTANCES

A. Pigments (pale to lemon yellow, red or orange)

   1 K+ purple or dark red   Anthraquinones and terphenylquinones.

   2 K−   Tetronic acid derivatives, usnic acid.

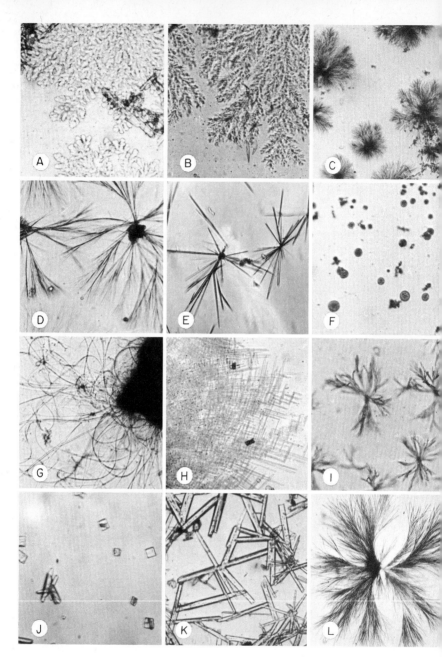

Plate 14 Recrystallized lichen substances. (A) Caperatic acid in G.E. (B) Proto-lichesterinic acid in G.E. (C) Lecanoric acid in G.A.W. (D) Cryptochlorophaeic acid in G.A.W. (E) Grayanic acid in G.E. (F) Gyrophoric acid in G.E. (G) Olivetoric acid in G.A.W. (H) Divaricatic acid in G.E. (I) Physodic acid in G.A.W. (J) Barbatic acid in G.E. (K) Usnic acid in G.E. (L) Psoromic acid in G.E.

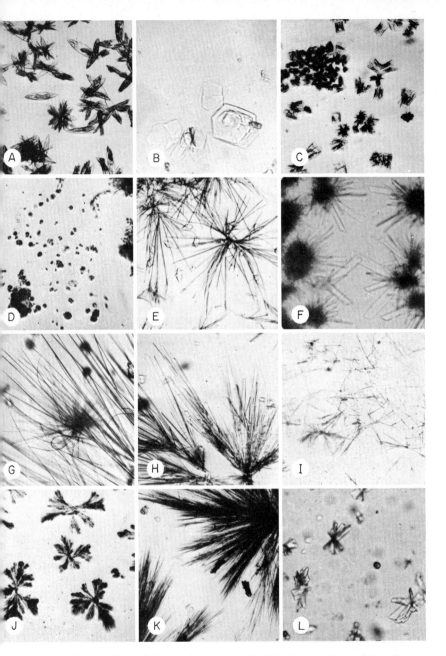

Plate 15 Recrystallized lichen substances. (A) Salacinic acid in G.A.o-T. (B) Stictic acid in G.A.o-T. (C) Norstictic acid in G.A.o-T. (D) Protocetraric acid in G.A.o-T. (E) Zeorin (prisms) and atranorin (needles) in G.A.o-T. (F) Barbatic acid in G.A.o-T. (G) Perlatolic acid in G.A.W. (H) Alectoronic acid in G.E. (I) Tenuiorin in G.A.o-T. (J) Evernic acid in G.E. (K) Thamnolic acid in G.A.An. (L) Alectoronic acid in G.A.W.

B. Colourless substances

1 K+, P+ yellow, orange or red	Atranorin, baeomycic acid, barbatolic acid, chloratranorin, norstictic acid, physodalic acid, salacinic acid, stictic acid, thamnolic acid.
2 K— or K+ brownish, P+ red or yellow.	Fumarprotocetraric acid, protocetraric acid, psoromic acid, pannarin.
3 K—, C+ red or rose	Anziaic acid, erythrin, gyrophoric acid, hiascic acid, lecanoric acid, methyl 3,5-dichlorolecanorate, olivetoric acid.
4 K—, C+ green	Didymic acid, strepsilin.
5 K—, C—, KC+ red or rose	Alectoronic acid, α-collatolic acid, glomelliferic acid, lobaric acid, norlobaridon, physodic acid, ramalinolic acid.

The remaining lichen substances do not react with K, C or P.

Alectoronic acid. Gummy residue from acetone; long colourless lamellae in G.E. (Plate 15H), fan-shaped lamellae in G.A.W. (Plate 15L). This substance is usually accompanied by α-collatolic acid. *Parmelia arnoldii, P. incurva.*

Anthraquinones. Chromatography used for identification. *Caloplaca* spp., *Xanthoria parietina*, red-fruited *Cladonias.*

Atranorin. Usually in the upper cortex and often accompanied by other depsides or depsidones; curved yellow needles in G.A.*o*-T. (Plate 15E), short rhombic needles in G.E. *Parmelia perlata, Physcia stellaris.*

Barbatic acid. Colourless prisms or lamellae in G.E. (Plate 14J), long colourless lamellae in G.A.*o*-T. (Plate 15F). *Cladonia bacillaris, C. floerkeana, Parmelia laevigata.*

Caperatic acid. Globular branched masses in G.E. (Plate 14A). *Parmelia caperata.*

Cryptochlorophaeic acid. Long colourless curved needles in G.A.W. (Plate 14D) and in G.E. *Cladonia cryptochlorophaea.*

Didymic acid. Small clusters of colourless rhombic needles in G.A.W. *Cladonia bacillaris.*

Divaricatic acid. Criss-cross needles in G.E. (Plate 14H). *Haematomma ventosum, Parmeliopsis ambigua.*

Evernic acid. Bushy colourless needle clusters in G.E. (Plate 15J). *Evernia prunastri, Parmelia taylorensis.*

Fumarprotocetraric acid. Fascicles of short straight yellow needles in G.A.*o*-T.; crystal tests often unsatisfactory and chromatography advisable, although the K−, P+ instant brick-red colour reaction is sufficient for identification. *Cetraria islandica, Cladonia rangiferina.*

Grayanic acid. Acetone residue deposited as coarse straight needles which recrystallize in a similar form in G.E. (Plate 14E). *Cladonia grayi.*

Gyrophoric acid. Colourless warts or granules in G.E. (Plate 14F). *Parmelia borreri, P. revoluta, Umbilicaria* spp.

Lecanoric acid. Bushy clusters of colourless curved needles in G.A.W. (Plate 14C) and G.E. *Diploschistes scruposus, Parmelia subaurifera, P. tiliacea.*

Lobaric acid. Dendritic clusters of colourless thick rhombic needles in G.E., easily confused with atranorin; strongly white fluorescent in ultraviolet light. *Stereocaulon paschale.*

Norstictic acid. Pale yellow four-angled lamellae in G.A.*o*-T. (Plate 15C), short, red, acicular needles in KOH—K_2CO_3 solution. *Parmelia acetabulum.*

Olivetoric acid. Acetone residue gummy; long thin curved needles in G.A.W. (Plate 14G). *Cetrelia olivetorum, Pseudevernia furfuracea* (chemical strain).

Perlatolic acid. Acetone residue gummy; fascicles of long radiating needles in G.A.W. (Plate 15G) and in G.A.Q. *Cetrelia cetrarioides.*

Physodic acid. Short somewhat flattened and curved colourless needles in G.A.W. (Plate 14I). *Hypogymnia physodes, Pseudevernia furfuracea.*

Protocetraric acid. Small yellow warts or balls in G.A.*o*-T. (Plate 15D), intergrading with fumarprotocetraric; chromatography desirable. *Parmelia caperata.*

Protolichesterinic acid. Feathery colourless lamellae in G.E. (Plate 14B). *Parmelia reddenda, Platismatia glauca, Pycnothelia papillaria.*

Psoromic acid. Fascicles of curved colourless needles in G.E. (Plate 14L); strongly yellow fluorescent on chromatograms in ammonia under U.V. light. *Cladonia alpicola, Rhizocarpon geographicum.*

Salacinic acid. Yellow, boat-shaped lamellae in G.A.*o*-T. (Plate 15A), sheaves of dark-red curved needles in KOH—K_2CO_3. *Parmelia reticulata, P. saxatilis.*

Squamatic acid. Solitary or aggregated colourless short prisms in G.E.; strong white fluorescence in U.V. light. *Cladonia squamosa.*

Stictic acid. Colourless hexagonal lamellae in G.A.*o*-T. (Plate 15B). *Lobaria pulmonaria, Menegazzia terebrata, Parmelia conspersa, P. perlata.*

Tenuiorin. Colourless fine needles in G.A.*o*-T. (Plate 15I). *Lobaria linita, Peltigera polydactyla.*

Tetronic acid derivatives. Identified with chromatography. *Candelaria concolor, Cetraria pinastri, Lepraria chlorina, Rhizocarpon geographicum.*

Thamnolic acid. Effervescence and formation of yellowish, fine, straight needles in G.A.An. (Plate 15K). *Parmeliopsis aleurites, Usnea florida.*

Usnic acid. Yellow straight needles in G.E. (Plate 14K). *Evernia prunastri, Parmelia conspersa, Ramalina* spp., *Usnea* spp.

Zeorin. Colourless hexagonal double prisms in G.A.*o*-T. (Plate 15E) and G.A.An. *Anaptychia speciosa, Cladonia deformis.*

PARTITION CHROMATOGRAPHY   As more and more workers use Asahina's tests, certain limitations are becoming evident. Some substances do not form crystals at all; certain ones, such as fumarprotocetraric acid, are difficult to crystallize; and others are so modified by impurities or interference from other substances that they cannot be positively identified. Consequently, perhaps 10% of the microchemical tests do not give conclusive results in routine testing. This gap has now been filled in large measure by paper chromatography. Beginning with the studies of the Swedish chemist Wachtmeister in 1951, chromatography of lichen substances has rapidly evolved into a standardized technique with many

Table 8.1 R_F values for the common lichen acids in various solvents (1 = *n*-butanol-NH_4OH; 2 = *n*-butanol-H_2O with phosphate-buffered (pH 9.0) paper; 3 = the same as 2 except with phosphate-buffered (pH 11.7) paper; 4 = *n*-butanol-acetone-water).[120]

	1	2	3	4
Usnic acid	0.92	0.88	—	—
Vulpinic acid	0.88	0.83	0.88	—
Calycin	0.73	—	—	—
Pulvic anhydride	0.22	—	—	—
Gyrophoric acid	0.26	0.70	0.32	—
Lecanoric acid	0.32	0.75	0.48	—
Olivetoric acid	0.89	0.92	0.84	1.00
Alectoronic acid	0.50	—	—	—
Lobaric acid	0.80	—	—	—
Physodic acid	0.77	—	—	—
Atranorin	0.73	0.95	0.47	1.00
Fumarprotocetraric acid	—	0.08	0.01	0.37
Norstictic acid	—	0.68	0.27	0.70
Protocetraric acid	—	0.50	—	0.56
Psoromic acid	—	0.69	0.47	0.68
Salacinic acid	—	0.44	0.01	0.59
Stictic acid	—	0.74	0.55	0.61
Thamnolic acid	0.07	0.08	0.01	0.32
Divaricatic acid	0.90	—	—	—
Evernic acid	0.64	—	—	—
Perlatolic acid	0.96	—	—	—
Squamatic acid	0.18	0.16	0.18	0.30

advantages over slide tests.[274] It is especially useful in separating tetronic acid derivatives and other pigments as well as several P+ red depsidones previously inseparable by crystal tests. Some typical R_F values are given in Table 8.1.

A more recent chromatographic technique, thin-layer chromatography, utilizes a thin layer of silica sprayed on a glass plate or plastic sheet.[36] Although a considerably more expensive technique than that of paper chromatography if pre-coated plates are purchased commercially, it has been widely adopted by lichen chemists and systematists and is now fairly standardized.[66A, 230A] The usual solvents (v/v) are benzene-dioxane-acetic acid (180:45:5), hexane-ether-formic acid (130:80:20), and toluene-acetic acid (200:30). Spots are 'visualized' by spraying with a 10% solution of sulphuric acid and heating the plates for 15 minutes at 100°C. Separation and resolution of the spots are far better than in paper chromatography.

Gas chromatography has not been used routinely to identify lichen substances, but it appears to be useful in analysing such difficult groups as the triterpenoids and fatty acids, which have not yet been satisfactorily studied with other microchemical tests. In this technique the lichen substances are volatilized and the time of passage of the compound and any thermal decomposition products plotted. A further refinement of this technique being used in chemical laboratories is high-pressure liquid chromatography, as reported by C. F. Culberson.[65B] The substances to be investigated are injected in the column as a solution, by-passing the problems of volatilizing the samples. As with gas chromatography, this technique provides information on the actual concentration of the lichen substances.

MASS SPECTROMETRY   This technique is widely employed in organic chemistry to identify microgram quantities of compounds. It has been used successfully with lichens to derive mass spectra for various pigmented substances (usnic acid, tetronic acid derivatives, quinones, etc.).[230B]

SENSITIVITY OF MICROCHEMICAL TESTS   The sensitivity of microchemical tests is becoming a more critical problem as the tests gain in popularity. If atranorin, for example, occurs in quantities of less than 0.9 μg in an acetone extract, it will not be detected in a crystallizing test. Chromatography is substantially more sensitive, but even here quantities below 0.12 μg will not be detected. For usnic acid the minimal quantities are 2.0 μg and 10.0 μg respectively.[63] These minima do not take into account variables such as interference by impurities that are known to lessen the sensitivity of microchemical tests.

9

Biochemical Systematics

Biochemical systematics is the application of biochemistry to taxonomic problems. Differences in chemical constituents of plants are used to complement or even supplant morphological characters. A large storehouse of information on plant chemistry has accumulated in the past century through the efforts of pharmacologists and biochemists, but it is only in the past 15 years that the development of partition chromatography has brought a rapid and sure means of identifying plant products within the reach of taxonomists and physiologists. In higher plants there are broad correlations at the order, family and genus level between the occurrence of alkaloids, steroids, essential oils, phenols and other compounds and systems of classification based only on morphology but, in general, biochemical systematics is poorly developed at the species level.[253A]

The use of chemistry in lichen taxonomy began when colour differences came to be accepted as generic and specific criteria. *Candelaria* and *Xanthoria*, for example, are old generic names for yellow and orange lichens that contain tetronic acid or anthraquinone derivatives, but are otherwise closely related to unpigmented genera. At the specific level, *Parmeliopsis ambigua*, a yellow lichen with usnic acid, is often separated from grey *P. hyperopta* which is morphologically identical but lacks usnic acid. A parallel case is *Parmelia centrifuga* (yellow) and *P. aleuriza* (mineral grey), which appear to be colour variants of a single population.

Most lichen substances, however, are colourless and can be detected only by indirect means. It was Nylander in the 1860s who devised the first practical method of identifying these compounds by simply daubing potassium hydroxide or calcium hypochlorite on the thallus to elicit colour changes.[208] Colour tests are now utilized by the majority of lichenologists as aids in identification, and in no other plant group has chemistry been used so successfully as in lichens. The uniqueness of lichen substances has immensely simplified the adoption of biochemical systematics. The total number of such substances is only about 80 and their molecular structures are almost all known. Positive identification is now possible with microchemical tests which require relatively simple techniques. Hence lichenologists are in the enviable position of being able to make routine use of unique compounds of known structure that often occur in high concentration. In contrast, botanists studying phanerogams are faced with hundreds of diverse, often unidentified, compounds when they attempt to make biochemical analyses.

VARIATION IN CHEMICAL COMPOSITION

Of the hundreds of different lichens chemically tested in the past century, the majority have constant chemical composition regardless of the substrate or geographic origin of the specimens tested. Thus *Parmelia perlata*, a pantemperate species in the northern and southern hemispheres, always contains atranorin and stictic acid.[124] *P. caperata*, another pantemperate species, invariably produces usnic, protocetraric and caperatic acids, and the common arctic reindeer moss, *Cladonia rangiferina*, always contains fumarprotocetraric acid. This uniformity is indeed fortunate and has proved to be a useful, though not indispensable, criterion for recognizing these species.

Other lichen populations that from all appearances are morphologically indistinguishable have variable chemical composition, and these variants are called chemical strains or, less commonly, races or phases. Nylander can probably be credited with the first discovery of chemical variation in 1866 when he identified among specimens of the common European lichen *Cetrelia cetrarioides* some plants reacting C+ red (olivetoric acid) and some C− (perlatolic acid) in the medulla. Similar patterns of chemical variation have since been reported in some of our commonest lichens; the species listed in Table 9.1 serve as representative examples of these chemical strains.

Number of chemical strains

Unfortunately it is not possible at this time to make an accurate summary of all lichens that have chemical strains. Only fairly recently published taxonomic revisions which employ acceptable chemical techniques can form the basis for an inventory. A number of valuable monographs, including those published before Asahina's introduction of microchemical methods in 1936, are unusable for this purpose because the biochemical features of the species are not considered. In other instances, chemists have correctly identified the substances, but there is grave doubt about the taxonomic determinations of the specimens they used for extraction. Furthermore, a few populations that have been regarded as strains are being shown by more careful study to be morphologically different, thus removing them from the ranks of true strains.

If, nevertheless, we tabulate the obvious examples of chemical strains as reported in recent literature, we can get a fair approximation of the frequency of chemical variation. The following genera are known to include species that have chemical strains; most, in fact, are rather rich in strain formation: *Alectoria*,[22] *Anaptychia*,[162] *Anzia*,[23,164] *Cetraria*,[122] *Cladonia*,[28,75,89] *Cornicularia*,[27] *Graphina*,[284] *Graphis*,[284] *Haematomma*,[32,72] *Oropogon*,[22,29] *Parmelia*,[70,73,115,123,125] *Ramalina*,[26,65] *Rhizocarpon*,[226] *Rinodina*,[109] *Stereocaulon*,[165] *Thamnolia*[233] and *Usnea*.[30,121]

Within this series of 17 genera there are by actual count 99 morphologically distinct populations or species in the classical sense and 141 chemical variants, for a total of 240 discrete chemical entities. It is estimated that 10–20% of the species now recognized in these genera are chemical strains. Other common foliose and crustose genera also seem to have numerous strains but only cursory investigations of them have been made. Gelatinous lichens as a group do not contain distinctive chemical substances and cannot be expected to have strains.

Patterns of strain formation

The combinations of acids in strains fall into several patterns. The most frequent, accounting for nearly 75% of known strains, is a complete substitution of the principal or diagnostic components, as shown below:

Strain 1	Strain 2	Strain 3	Strain n
Acid A	Acid B	Acid C	Acid X

The principal component is very often accompanied by a constant accessory component, usually atranorin or usnic acid, but these are omitted for purposes of simplification in this and the following patterns. The diagnostic acids rarely, if ever, occur together in the same plant and may have different molecular structure, though they are not necessarily biogenetically unrelated. Common examples are *Rinodina oreina*, *Pseudevernia furfuracea* and *Cetraria ciliaris* (Table 9.1), as well as *Parmelia bolliana* (lecanoric or protolichesterinic acid) and the *Cladonia chlorophaea* group (cryptochlorophaeic, merochlorophaeic, grayanic or fumarprotocetraric acid alone).

A modified and somewhat more complex combination characterizing *Thamnolia vermicularis* may be represented as follows:

Table 9.1 Chemical strains of some selected lichens. Only diagnostic substances are listed.

	Strain 1	Strain 2	Strain 3
Anaptychia diademata[162]	atranorin, zeorin	atranorin, zeorin, salacinic acid	—
Cetraria ciliaris[122]	olivetoric acid	alectoronic acid	protolichesterinic acid
C. islandica[24]	fumarprotocetraric, protolichesterinic acids	protolichesterinic acid	—
Graphis caesiella[284]	norstictic acid	stictic acid	protocetraric acid
Haematomma puniceum[32]	atranorin and unknown	imbricaric acid	atranorin only
Oropogon loxensis[29]	P + unknown	psoromic acid	P – unknown
Parmelia cetrata[127]	salacinic acid	protolichesterinic acid	—
Pseudevernia furfuracea[115]	physodic acid	olivetoric acid	lecanoric acid
Ramalina sekika[26]	sekikaic acid	obtusatic, evernic acids	—
Rinodina oreina[109]	fumarprotocetraric, usnic acids	gyrophoric, usnic acids	usnic acid
Usnea strigosa[121]	norstictic acid	psoromic acid	no substances

Strain 1	Strain 2
Acid A	Acid B + C

Another pattern, where one strain contains no diagnostic substances at all, has been called an acid-deficient strain or inactive phase:

Strain 1	Strain 2
Acid A	No acids

Usnea antarctica (fumarprotocetraric acid or no acid) is an example of this type.[170] Ignoring the constant components divaricatic and usnic acids, we can see the same pattern in *Haematomma ventosum*, which may contain thamnolic acid or lack it.[72]

A fourth pattern, which accounts for about 20% of known strains, may be described as additive. The two strains share a diagnostic substance but one strain has an additional substance with variable frequency in the population, according to this pattern:

Strain 1	Strain 2
Acid A	Acid A + B

Rhizocarpon cinereovirens[54] and *Parmelia conspersa* (stictic acid or stictic and norstictic acids) are examples of this type.

A final pattern is a combination of substitution and additive strains:

Strain 1	Strain 2	Strain 3
Acid A	Acid B	Acid A + B

This is known in *Lobaria pulmonaria* (stictic acid or norstictic acid or stictic and norstictic acids) and in *Anzia opuntiella* (divaricatic acid or sekikaic acid or divaricatic and sekikaic acids).[164]

As one would expect, a number of strains are quite complex and cannot be accommodated within the patterns listed above. The common fruticose European lichen *Ramalina siliquosa*, for example, has these combinations of acids in six strains:[65] (1) Hypoprotocetraric and usnic; (2) norstictic, stictic and usnic; (3) norstictic and usnic; (4) protocetraric and usnic; (5) salacinic and usnic; and (6) usnic alone.

Lichen substances in chemical strains

If we summarize the kinds of lichen substances found in the 240 strains mentioned above, almost all of the various groups are represented, but the depsidones reacting with P have overriding numerical frequency. Of the total, 33 strains contain salacinic acid, 24 fumarprotocetraric acid, 22 norstictic acid, 21 stictic acid, 12 fatty acids, 10 squamatic acid, 10 thamnolic acid, 8 protocetraric acid, 8 divaricatic acid, 8 barbatic acid, 8 cryptochlorophaeic acid, with 24 other substances accounting for the remaining 76 strains. Some strains contain unrelated substances, as lecanoric acid (a depside) or protolichesterinic acid (a fatty acid) in *Parmelia bolliana*, whereas others contain substances such as stictic and norstictic acids or physodic and olivetoric which differ in minor structural features, as length or oxidation of side chains. From a purely chemical viewpoint, it is probably justifiable to recognize every chemical variant as distinct, but in practice taxonomists assign different weight to them. Since atranorin and usnic acid, for example, are constant components in many genera, they are usually thought to have no significance in biochemical taxonomy. It is still too early, however, to decide on the relative importance of structural differences of lichen substances.

BIOGENESIS OF CHEMICAL STRAINS

The basic patterns of strain formation can be interpreted as metabolic shunts where the end-products are stable crystalline lichen substances deposited on the surface of the hyphae. While there is exceedingly little information on the biogenesis of these substances, the following two theoretical pathways could lead to strains:

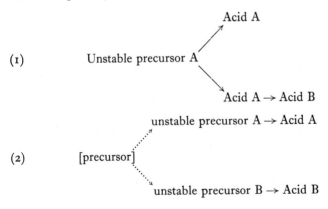

Strains with pairs of homologous depside-depsidones, while rare, substantiate the first pathway. *Pseudevernia furfuracea* in Europe and *Cornicularia satoana* in Asia, for example, contain either olivetoric or physodic acid.[27,115] Physodic acid, a depsidone, can theoretically be derived from the depside olivetoric acid by dehydrogenation (Fig. 9.1). Thus all plants would appear to synthesize an unstable precursor that is converted into olivetoric acid; this may undergo no further change, but a part of the population with the necessary enzymes is able to convert olivetoric acid into physodic acid. That this conversion is not always carried to completion is evident in the discovery by Chicita Culberson of thalli of *P. furfuracea* in Spain that contain both olivetoric and physodic acid.[64] The rarity of the strain pattern, Acid A–Acid B–Acid AB, however, is convincing evidence of the efficiency of the conversion mechanism.

The acids in most strains do not have as close a biogenetical relationship as olivetoric and physodic acids. They differ frequently in the number of carbon atoms in the side chains, a number that is determined in very early stages of biogenesis as compared with the dehydrogenative conversion. In these examples the second pathway (2) above is more likely to occur, with the strains derived from different precursors by separate routes. Recent work by Shibata on the possible biogenesis of strains in the *Cladonia*

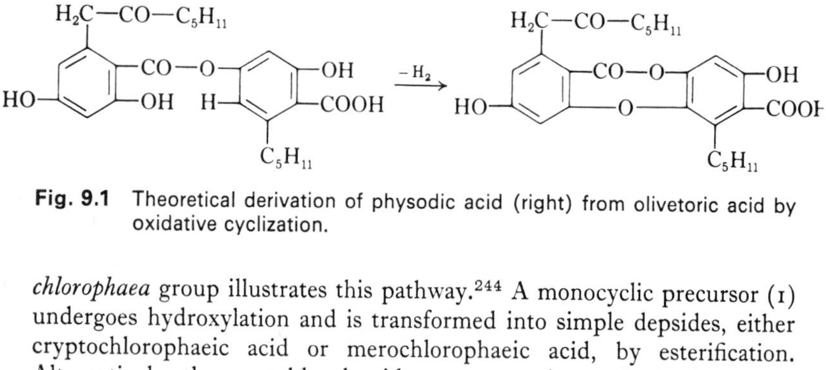

Fig. 9.1 Theoretical derivation of physodic acid (right) from olivetoric acid by oxidative cyclization.

chlorophaea group illustrates this pathway.[244] A monocyclic precursor (1) undergoes hydroxylation and is transformed into simple depsides, either cryptochlorophaeic acid or merochlorophaeic acid, by esterification. Alternatively the unstable depside precursor is synthesized without hydroxylation, but later undergoes oxidative cyclization to produce the depsidone grayanic acid (Fig. 9.2).

Runemark[226] has proposed a similar scheme to account for the strains in the *Rhizocarpon lindsayanum* group. The depside barbatic acid accumulates in one strain because of a block in the dehydrogenative mechanism, while other strains possessing this mechanism go on to form depsidones, stictic acid or the related derivatives norstictic or psoromic acid.

Factors that determine the direction of metabolic shunts in lichens are unknown. This problem is central to a final solution of the taxonomic and evolutionary significance of chemical strains. It is a reasonable assumption that strain formation is an expression of the genotype of the fungus and that enzymatic systems are primarily responsible for the biosynthetic transformations. This would be in line with what is already known about the metabolism of other fungi but there is no way at present to prove it for lichens. Zopf[289] in his studies of the *Pseudevernia* group left the impression that the chemical contents of the plants varied according to the bark substrate, birch, pine, etc., but his data are insufficient to draw any such conclusions. Mass sample studies of *Parmelia bolliana* on two different oaks in central United States have failed to show any possible correlation with substrate or micro-environment,[120] and strains of *Cetraria ciliaris* are

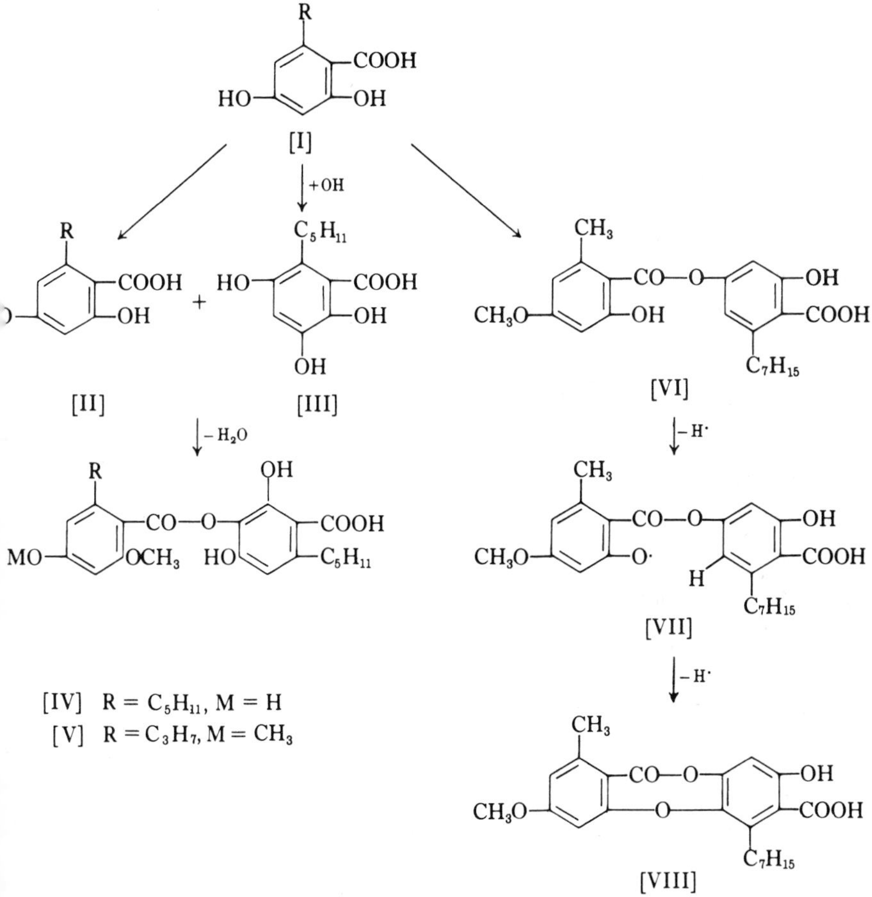

Fig. 9.2 Hypothetical biogenesis of lichen acids in the *Cladonia chlorophaea* group. (I) = a monocyclic phenylcarboxylic acid, (III) = hydroxylation product, (II) and (III) forming a depside by esterification, (IV) = cryptochlorophaeic acid, (V) = merochlorophaeic acid. (VI) = a depside derived from self-esterification of two molecules of (I), (VII) = intermediate stage in oxidative cyclization, and (VIII) = grayanic acid. (After Shibata and Hsüch-Ching[244])

randomly distributed on trees as different as *Pinus strobus* and *Quercus alba*.[122] Environmental factors, however, should not be wholly disregarded as a possible cause of strain formation, since the quantity of the pigments usnic acid and parietin is known to vary with light intensity.[145]

GEOGRAPHICAL DISTRIBUTION OF STRAINS

The existence of chemical strains in a lichen population can be established theoretically on the basis of as few as two specimens, but full-scale investigations of the proportions of strains in a population and the patterns of distribution call for analysis of hundreds of specimens. Inasmuch as the need for mass sampling has only recently been appreciated and rapid microchemical identification become available, distributions of extremely few strains are known at this time. Examination of herbarium specimens alone is usually insufficient to detect differences in the behaviour of strains in micro-environments or ecotones where they overlap, and fairly intensive field studies must often be pursued independently of herbarium work. Time and cost factors in this type of investigation are so great that progress in determining distribution patterns of strains is regrettably slow.

Geographical distributions have been more or less completely worked out on a continental level for *Cetraria ciliaris*,[122] *Parmelia arnoldii*,[124] *P. bolliana*,[73] *P. cetrarioides*,[68] *P. cirrhata*,[127] *P. conspersa*,[123] *P. hababiana*,[124] *P. perforata*,[124] *P. plittii*,[123] *P. subsumpta*,[124] *P. subtinctoria*,[124] *Pseudevernia furfuracea*,[115] *Rinodina oreina*,[109] *Thamnolia vermicularis*[233] and *Usnea strigosa*.[121] Brief summaries of some of these species will serve to illustrate the kind of variation that has been uncovered. Most strains have bi- or multi-centric patterns that coincide with phytogeographic areas already recognized for higher plants, but it is premature to speculate on the significance of these patterns other than to say that they appear to support the idea of genic origin for the strains.

Parmelia bolliana

Parmelia bolliana is a common lichen on oak trees in the prairie regions and deciduous forests of North America. The protolichesterinic acid strain (C−) has a broad distribution from New England to Texas and Minnesota (Fig. 9.3). The lecanoric acid strain (C+) is strongly bounded along the northern and eastern limits of its range.[73] Where the two strains overlap in Kansas, Oklahoma and Texas, there is a complex mosaic of populations in which the proportion of plants with lecanoric acid varies from 10–90%. The abundance of apothecia and lack of the usual vegetative diaspores suggest sexual reproduction by spores, but the life cycle of this lichen is unknown.

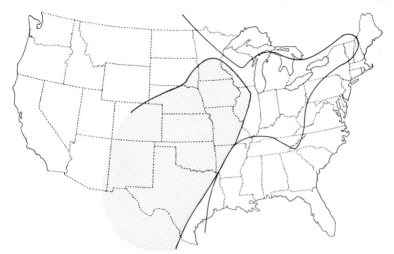

Fig. 9.3 Distribution of the strains of *Parmelia bolliana* in the United States: protolichesterinic acid (no shading) and lecanoric acid (stippled). (After Culberson and Culberson[73])

Parmelia cirrhata

Parmelia cirrhata, a common foliose montane species in tropical America and Asia, normally contains salacinic acid with or without protolichesterinic acid. In Mexico and Central America, however, there are two additional populations, one with norstictic acid (Fig. 9.4) and one with gyrophoric and protolichesterinic acids, which together account for 60% of the total population. There is a parallel pattern of strains in the closely allied species *P. nepalensis*. It is difficult to understand why strains have evolved so rapidly here but not in Asia, where both species are also very abundant.[128A]

Fig. 9.4 *Parmelia cirrhata* (natural size). (Drawing by L. Anderson)

Parmelia conspersa

The saxicolous *Parmelia conspersa* group with a pale lower surface has an especially complex pattern of strain distribution in North America (Fig. 9.5). The strain with stictic acid (*P. plittii* Gyel.) (Plate 16C) is

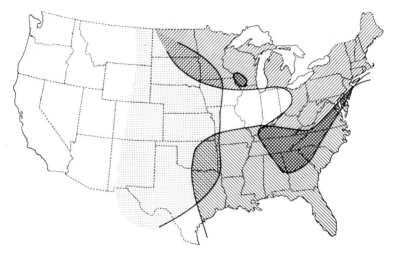

Fig. 9.5 Distribution of the strains of *Parmelia plittii* in the United States : stictic acid (diagonal lines sloping to left), salacinic acid (stippled area), and fumarprotocetraric acid (diagonal lines sloping to right). (Adapted from Hale[123])

wide ranging, although probably absent or extremely rare in the Ozark region. The obvious centre of origin for this strain is the Appalachian Mountains. The salacinic acid strain is restricted to central and western United States, and may have originated in Mexico. The fumarprotocetraric

Table 9.2 Percentage of chemical strains in population samples of *Parmelia plittii* in four areas in central United States. Within each pair the sample areas are separated by 300–500 kilometres.[123]

	Sample size	Salacinic acid	Fumarproto-cetraric acid	Stictic acid
North-eastern Minnesota	437	1	0	99
South-western Minnesota	548	81	0	19
Western Arkansas–eastern Oklahoma	87	5	95	0
Oklahoma–Kansas–Texas	104	99	1	0

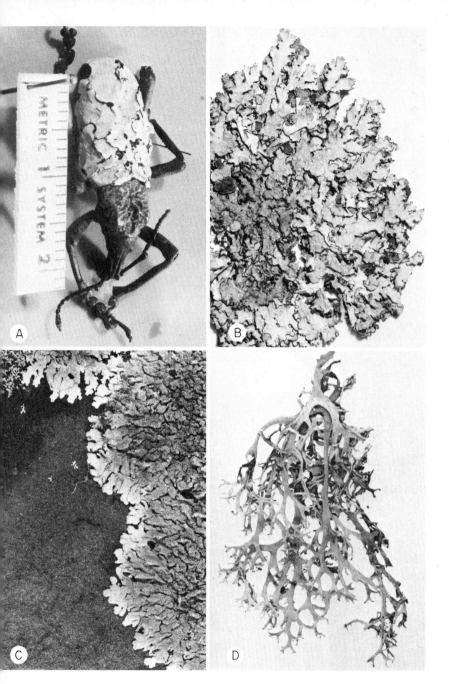

Plate 16 (A) The weevil *Gymnopholus lichenifer* from New Guinea covered with *Parmelia reticulata*. (B) Thallus of *Cetraria ciliaris* (olivetoric acid strain) (× 2). (C) *Parmelia plittii* (stictic acid strain) (× 2). (D) *Pseudevernia furfuracea* (lecanoric acid strain) (× 2). (A by J. L. Gressitt)

acid strain, a pantropical lichen, has a southern distribution outside of the limits of glaciation but may have had a much broader distribution in pre-glacial times, since it occurs as a disjunct in the celebrated unglaciated refugium of south-western Wisconsin. The boundaries of these three strains are amazingly sharp, even in mass samples (Table 9.2).[114]

Pseudevernia furfuracea

Two strains of *Pseudevernia furfuracea* (Plate 16D) are restricted to Europe. The olivetoric acid strain (C+) has a frequency of about 80% in Great Britain and a gradual diminution in the various countries of Europe with reciprocal increase in the frequency of physodic acid (C−), until 100% of the plants in North Africa contain physodic acid (Table 9.3).

Table 9.3   Percentage of chemical strains in herbarium samples of *Pseudevernia furfuracea*.[115]

	Olivetoric acid	Physodic acid
Great Britain	80	20
Norway	67	33
Sweden	42	58
Finland	40	60
Denmark	36	64
Netherlands	27	73
Germany	23	77
Switzerland	35	65
Austria	38	62
France	18	82
North Africa	0	100

Plants with joint occurrence of olivetoric and physodic acids make up about 1% of a single population in Spain but their frequency elsewhere is un-known.[64] Populations in North America react uniformly C+, caused not by olivetoric acid as in Europe but by lecanoric acid.[115]

Rinodina oreina

The gyrophoric acid and fumarprotocetraric acid strains of *Rinodina oreina* (Plate 4D), a conspicuous crustose lichen, are common in eastern United States and the Great Lakes region.[109] The acid-deficient strain is largely confined to the central and western states. In Europe and the arctic areas, the acid-deficient and gyrophoric acid strains are much more common than the fumarprotocetraric acid strain. This lichen is suspected to reproduce at least in part sexually by spores.

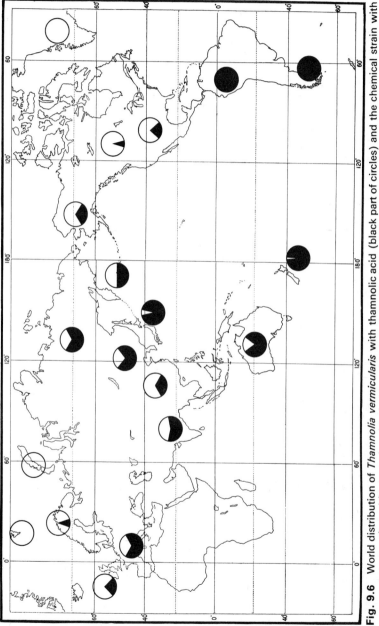

Fig. 9.6 World distribution of *Thamnolia vermicularis* with thamnolic acid (black part of circles) and the chemical strain with squamatic and baeomycic acids (white part of circles). (From Sato[233])

Thamnolia vermicularis

Thamnolia vermicularis, a common arctic-alpine fruticose soil lichen (Fig. 1.8b), has two strains, one with thamnolic acid and one with squamatic and baeomycic acids. The former is most abundant in the southern hemisphere, the latter in the northern hemisphere (Fig. 9.6).[233] Mass sample studies by Sato in Japan show a consistent increase in the percentage of thamnolic acid-containing plants moving southward from Hokkaido to central Honshu. The lack of apothecia in *T. vermicularis* seems to rule out any genotypic variation in strain formation that might be attributed to sexuality.

A few chemically differentiated races of lichens which have overlapping boundaries are distributed in different microhabitats. Culberson and Culberson[73A] call this phenomenon habitat selection. They clearly demonstrated this in three chemical populations of *Ramalina siliquosa* occurring together in the littoral zone in North Wales. Plants with stictic acid occupied the lowermost exposed zone, those with hypoprotocetraric acid the upper less exposed zone, and those with norstictic acid the intermediate zone. Preliminary studies of the *Cladonia chlorophaea* complex in Minnesota and Michigan, where four chemical populations converge, have shown that each race occupies a different habitat (soil, base of pine trees, etc.) to a statistically significant degree.[282A]

NOMENCLATURE OF CHEMICAL STRAINS

When Nylander recognized the C+ and C− populations of *Parmelia cetrarioides* in Europe, he called them distinct species, *P. olivetorum* and *P. cetrarioides*, respectively. Obviously far in advance of his contemporaries in the utilization and acceptance of colour tests as taxonomic criteria, he was nevertheless opposed by most lichenologists. Colour tests were slowly adopted, it is true, but only as aids in identification. *P. olivetorum*, if recognized at all, was called *P. cetrarioides* var. *rubescens*, and when later workers discovered other chemical strains, they usually preferred to call them forms or varieties of a single species.

The chemist Zopf[289] made several excursions into taxonomy, separating *Pseudevernia furfuracea* into three chemical species, *P. isidiophora* (C−), *P. furfuracea* (C−) and *P. olivetorina* (C+ red), but few taxonomists followed him. Indeed, chemistry seemed destined for a subordinate role in lichen taxonomy until 1929, when the Hungarian lichenologist Gyelnik began to describe colour-tested variants as distinct species so rapidly and recklessly that he incurred the wrath of nearly all lichenologists. This unfortunate chapter in lichen history was partially redeemed by the work of Asahina and his associates who, on the one hand, introduced microchemical tests and placed colour reactions on a more exact scientific basis and, on the other, brought some semblance of order into the taxonomic treatment of chemical strains.

In 1937 Asahina[24] propounded the working hypothesis that chemically different populations be recognized as distinct species. He proceeded to describe many 'chemical species' but in a more logical fashion than Gyelnik and on the basis of exacting chemical investigations. Duvigneaud,[83] a Belgian lichenologist, published a particularly lucid account of biochemical taxonomy in lichens in 1939, defending and adding support to Asahina's hypothesis. Evans[89] and Dahl[75] adopted chemical features as a species character in *Cladonia* and did much to unravel the taxonomy of this difficult genus. Recognition of chemical species is also gaining acceptance in the large foliose genus *Parmelia*.[124] It is no exaggeration to say that strain formation is turning out to be a valuable taxonomic character in almost every group that has been properly studied.

Most lichenologists, however, are still far from convinced that chemical characters are valid specific or infraspecific criteria. Lamb[165] proposed an alternative solution to this problem by calling each chemical variant a numbered 'chemical strain' of a parent species. Thus the single species *C. chlorophaea* is considered to have three chemical strains, Ch. str. II (grayanic acid), Ch. str. III (cryptochlorophaeic acid) and Ch. str. IV (merochlorophaeic acid), in addition to the fumarprotocetraric acid-containing parent strain, Ch. str. I. This extra-nomenclatorial approach has considerable merit in enabling the taxonomist to recognize chemical variation on an informal basis without being distracted by or being forced to make nomenclatorial decisions.

If we choose to use chemical features at the species level, the following taxonomic disposition of some of the commoner chemical strains mentioned in this chapter is possible:

Cetraria ciliaris Ach. (olivetoric acid), *Cetraria halei* W. Culb. (alectoronic acid) and *C. orbata* Nyl. (protolichesterinic acid).

Haematomma ventosum (L.) Mass. (divaricatic, usnic and thamnolic acids) and *H. lapponicum* Räs. (divaricatic and usnic acids).

Parmelia bolliana Müll. Arg. (protolichesterinic acid) and *P. hypoleucites* Nyl. (lecanoric acid).

Parmelia cirrhata Fr. (salacinic acid), *P. imitata* Hale & Wirth (gyrophoric acid) and *P. neocirrhata* Hale & Wirth (norstictic acid).[128A]

Parmelia plittii Gyel. (stictic acid), *P. mexicana* Gyel. (salacinic acid) and *P. subramigera* Gyel. (fumarprotocetraric acid).

Pseudevernia furfuracea (L.) Zopf (physodic acid), *P. olivetorina* (Zopf) Zopf (olivetoric acid) and *P. consocians* (Vain.) Hale & W. Culb. (lecanoric acid).

Rinodina oreina (Ach.) Mass. (fumarprotocetraric acid), *R. hueana* Vain.
(C−), *R.* sp. (unnamed) (gyrophoric acid).

Thamnolia vermicularis (Sw.) Ach. (thamnolic acid) and *T. subuliformis*
(Ehrh.) W. Culb. (squamatic and baeomycic acids).

While many strains have not yet been described at species level, it is an
interesting aside that some have already been named inadvertently because
taxonomists were ignorant of the existence of another previously named
strain. When Du Rietz described *Parmelia arnoldii* from Europe in 1927,
he had no idea that a morphologically identical chemical variant containing
salacinic instead of alectoronic acid had already been described from Ohio
in 1899 as *P. margaritata* by Hue.[124] Conservative taxonomists have
already unwittingly described 15 of the 18 chemical variants in *Parmelia*
subgenus *Amphigymnia* as distinct species, although to a man they would
have opposed chemical taxonomy. We are therefore more often confronted
with the question of what should be done with chemical species already
described than with the propriety of describing new ones.

Subspecific rank for chemical strains has gained in popularity lately.
Asahina himself has rather arbitrarily used subspecies for naming strains
along with specific rank. Most recently the rank of chemovar.[261] has been
proposed, as *P. borreri* chemovar. *borreri* and *P. borreri* chemovar. *pseudo-
borreri*. Any subspecific rank, however, seems to be undesirable at this
stage in chemical taxonomy of lichens, since it relegates a possible taxonomic trait, chemical features, to infraspecific importance without giving
it suitable prior study.

While taxonomists have been preoccupied chiefly with problems on the
taxonomic rank of chemical variation, it is also possible to demonstrate
a significant degree of correlation between chemical characteristics and
higher systematic categories. O-methylation, for example, is correlated
with more complex morphological development in *Parmelia*.[127A] On the
order level it is obvious that the Lecanorales is the seat of most chemical
variation and many families in the order can be arranged in a more natural
sequence by analysing molecular structures and biochemical evolution of
the substances.[73C] Much work remains to be done, of course, since only
about 2300 of 15 000 known species have been assayed for chemical
features.

The field of biochemical systematics is in a state of flux, but its role will
undoubtedly grow in the future. In some cases, of course, chemical
characters will be shown to have no intrinsic taxonomic value, but in others
they will have considerable utility. New data on correlations with geography
and morphology will continue to strengthen the use of taxonomic rank for
strains. No other field in lichenology promises as many new discoveries of
basic importance.

IO

Classification and Taxonomy

Few botanists are aware that lichens comprise one of the largest groups of described fungi and account for the majority of inoperculate Ascomycetes,[14] although they were not even recognized as fungi before 1866. Linnaeus classified lichens with the algae and used growth form alone to arrange them in nine series. Later attempts at internal classification were based on external differences in the shape of ascocarps. In 1831 the great Swedish mycologist Elias Fries, for example, divided lichens into two orders, one with open ascocarps (apothecia and mazaedia), the other with closed ascocarps (Pyrenomycetes, *Pertusaria*). Microscopic characters and ontogeny of the ascocarps were not taken into account.

SYSTEMS OF LICHEN CLASSIFICATION

The Finnish lichenologist Vainio in 1890 proposed a complete classification of lichens on the premise that lichens do not form a separate biological group but should be distributed among the fungi.[272] Fink[91] held the same view but made little headway in convincing his fellow botanists. In a poll made in 1911, he was discouraged to learn that 83% of them favoured keeping lichens as a separate group of plants. Vainio's system naturally benefited from the great advances in lichen morphology made in the latter half of the nineteenth century and the families and orders were grouped along more natural lines than in the scheme of Fries. Two main divisions were recognized, the Discomycetes, including the subdivisions Cyclocarpeae (apothecia present), Graphideae (lirelliform apothecia present), and Coniocarpeae (mazaedia present), and the Pyrenomycetes (perithecia present). Although followed in broad outline as recently as 1953 by Watson[279] in his *Census Catalogue of British Lichens*, Vainio's system was soon overshadowed by one proposed by Zahlbruckner and given wide circulation by its publication in the important *Natürlichen Pflanzenfamilien* of Engler and Prantl,[285] a standard reference for plant classification, and by its use as a framework for arranging species in the indispensable index to all lichen names, Zahlbruckner's *Catalogus Lichenum Universalis*.[286]

Zahlbruckner drew heavily on the researches of his contemporary, Reinke, who published a series of detailed studies on lichen morphology and a classification system in 1896.[222] Reinke unequivocally stated his belief that 'lichens are a special class different from and contrasting with fungi', in spite of the fact that he enumerated a number of genera transitional between lichens and fungi. Reinke's classification may be briefly summarized as follows:

Class Lichenes
1 Subclass Coniocarpi (ascocarp a mazaedium)
2 Subclass Discocarpi (ascocarp an apothecium)
   (a) Series Grammophori (apothecia lirelliform)
   (b) Series Lecideales (apothecia lecideine)
   (c) Series Parmeliales (apothecia lecanorine)
   (d) Series Cyanophili (lichens with blue-green algae, including the Peltigeraceae and Stictaceae)
3 Subclass Pyrenocarpi (ascocarp a perithecium)

Zahlbruckner rearranged these categories, assigned them different ranks, and added lichenized Basidiomycetes as follows:

A. Subclass Ascolichenes (Ascomycetes)
   Series Pyrenocarpeae (Pyrenomycetes)
   Series Gymnocarpeae (ascocarps open)
      Subseries Coniocarpineae (ascocarp a mazaedium)
      Subseries Graphidineae (apothecia lirelliform)
      Subseries Cyclocarpineae (apothecia disciform)

B. Subclass Hymenolichenes (Basidiomycetes)

This system was followed with only slight modification by Fink[92] and other leading taxonomists and is presently accepted or at least in common use by most lichenologists.

INTEGRATION OF LICHENS AND FUNGI

Lichens can no longer be maintained as a separate group of plants since they consist of two discordant elements, fungi and algae. The thalli and fruiting bodies are predominantly fungal in structure, and it is only logical that fungi be classified by the same system whether they are symbiotic or not. No taxonomic importance should be assigned to the algae. The old lichen-oriented systems, no matter how elaborate, are therefore unacceptable and must be rejected. The only alternative is to look toward contemporary systems of classifying non-lichenized fungi for guidance in setting up a more natural scheme within which lichens can be accommodated.

Lichenologists have been slow to re-evaluate lichen classification in the light of studies of mycologists. A rational integration of lichens with fungi is highly desirable, but the complexity of both groups makes this a difficult goal to achieve.[80] With the small number of workers now engaged in morphological and taxonomic research, it may take another 20 years before lichens are as well known as fungi. Attempts such as that by Clements and Shear[62] to distribute lichens among fungi have done little to clarify the basic problems, because the authors, unfamiliar with lichens, compiled their data from unverified or outdated sources. Nannfeldt[204] laid the guidelines for classifying lichens with fungi by separating ascohymenial and ascolocular types, a scheme first put to practical use by Santesson[232] in his exhaustive revision of the crustose foliicolous lichens. New and rather involved systems proposed by Korf[160] and Tomaselli,[270] which include lichens, have received little support from mycologists or lichenologists. As long as the hymenial and ascal wall structure of so many lichens remains either unstudied or in dispute, integration of lichens with fungi is no more than an unsatisfactory, often blind, juggling of families and orders that must be re-arranged as each new piece of morphological evidence is published.

There is no one 'right' way to classify lichens, but our present ignorance and hesitation to integrate them with fungi do not lessen the need for some kind of arrangement, even if subject to change in the future. It is not unreasonable to adopt a relatively conservative system such as that used by Dennis[80] for the British Ascomycetes. It is based on Nannfeldt's original divisions as modified by Luttrell, as follows:

Class Ascomycetes

1 Subclass Ascomycetidae
   Order Lecanorales
   Order Sphaeriales
   Order Caliciales
2 Subclass Loculoascomycetidae
   Order Pleosporales
   Order Hysteriales

FAMILIES AND GENERA OF LICHENS

The following list is a provisional arrangement of the major families and genera of lichens. It is not claimed to be exhaustive but does include most of the genera cited in James' checklist of British lichens[150] and Hale and Culberson's list of American lichens.[128] More detailed tabulations of genera may be consulted in Zahlbruckner[285] or in more recent monographic treatments. The families are ordered along a phylogenetic series from 'primitive' to 'advanced' development as generally accepted by systematists, although this sequence too will undoubtedly be subject to considerable revision in the future. Apothecial ontogeny is given only when exceptional. The thallus is stratified (heteromerous) and the phycobiont green (usually *Trentepohlia* or *Trebouxia*) unless otherwise mentioned.

Fruiting bodies containing asci:
CLASS ASCOMYCETES

Asci with one wall (unitunicate), regularly arranged
in a hymenium with free unbranched paraphyses:

SUBCLASS ASCOMYCETIDAE

Fruiting bodies apothecia:

Order Lecanorales

Lichinaceae.[139] Thallus gelatinous, granulose to filamentous, unstratified; phycobiont blue-green; apothecia arising from several contiguous ascogonia and adjacent vegetative hyphae, primary and secondary paraphyses present. Habitat: saxicolous in moist areas. Distribution: arctic to temperate. Genera: *Anema, Ephebe, Lichina, Lichinella, Phylliscum, Porocyphus, Synalissa, Thermutis, Zahlbrucknerella.*

Collemataceae.[78, 141] Thallus gelatinous, unstratified, granular, subfruticose or foliose; phycobiont *Nostoc*; apothecia derived entirely from the ascogone and associated stalk cells. Habitat: soil, rocks and trees in moist areas. Distribution: circumboreal, pan-temperate and montane-tropical. Genera: *Arctomia, Collema* (Plate 1C), *Leciophysma, Lemmopsis, Lempholemma, Leptogium, Physma.*

Heppiaceae. Thallus subfoliose to squamulose, unstratified; phycobiont *Scytonema*. Habitat: soil in semi-arid regions. Distribution: temperate and subtropical. Genus: *Heppia.*

Pannariaceae. Thallus squamulose or foliose, upper cortex well developed, lower cortex lacking but a conspicuous hypothallus often present; phycobiont blue-green. Habitat: soil and trees in wetter areas. Distribution: arctic to temperate and montane-tropical. Genera: *Erioderma* (Plate 7F), *Pannaria* (Plate 7D), *Parmeliella, Psoroma.*

Coccocarpiaceae.[139] Thallus typically foliose, stratified, with upper and lower cortex; phycobiont blue-green; apothecia lecideine. Habitat: rocks and trees. Distribution: temperate and tropical with highest development in tropical Asia and the Pacific region. Genera: *Coccocarpia, Spilonema.*

Peltigeraceae.[140] Thallus subfruticose to foliose, lower cortex poorly differentiated; phycobiont blue-green or green; apothecial development hemiangiocarpous. Habitat: trees, soil and humus. Distribution: circum-boreal and pan-temperate. Genera: *Hydrothyria, Koerberia, Massalongia, Nephroma, Peltigera* (Plate 8C), *Placynthium, Polychidium, Solorina.*

Stictaceae. Thallus foliose, often large, lower cortex often tomentose; phycobiont blue-green or green; cyphellae or pseudocyphellae common. Habitat: trees. Distribution: temperate and tropical with greatest abundance in the southern hemisphere. Genera: *Lobaria, Pseudocyphellaria, Sticta* (Plate 8A).

Graphidaceae.[284] Thallus crustose, stratified; phycobiont green; apothecia lirelliform. Habitat: trees. Distribution: subtropical and tropical with greatest development in south-eastern Asia. Genera: *Graphina, Graphis, Glyphis, Gyrostomum, Medusulina, Melaspilea, Phaeographina* (Plate 9C), *Phaeographis, Sarcographa, Xylographa.*

Thelotremataceae. Thallus crustose; apothecia round, more or less immersed, with a central columella, proper and thalline exciples well developed. Habitat: trees. Distribution: tropical and subtropical. Genera: *Leptotrema, Ocellularia, Phaeotrema, Thelotrema.*

Asterothyriaceae.[232] Thallus crustose; apothecia immersed. Habitat: leaves of evergreen trees. Distribution: tropical. Genus: *Gyalectidium.*

Gyalectaceae. Thallus crustose or filamentous; phycobiont often *Trentepohlia*; apothecia lecideine. Habitat: trees. Distribution: temperate and montane-tropical. Genera: *Coenogonium* (Plate 3A), *Dimerella, Gyalecta, Pachyphiale, Ramonia.*

Lecideaceae. Thallus crustose, rarely squamulose; phycobiont usually *Trebouxia*; apothecia lecideine. Habitat: rocks and trees. Distribution: arctic to temperate and montane-tropical. Genera: *Bacidia* (Plate 3B), *Bombyliospora, Byssoloma, Catillaria* (Plate 3C), *Lecidea* (Plate 2B; Plate 9B), *Lopadium, Megalospora, Mycoblastus, Phyllopsora, Protoblastenia, Rhizocarpon* (Plate 4A), *Tapellaria, Toninia.* This is the largest lichen family, *Lecidea* having over 1000 species.

Stereocaulaceae.[165] Thallus fruticose; apothecia lecideine. Habitat: rocks and soil. Distribution: arctic, temperate and montane-tropical. Genus: *Stereocaulon.*

Cladoniaceae. Primary thallus crustose or squamulose, secondary thallus (podetium) fruticose; apothecia lecideine. Habitat: rocks and soil. Distribution: arctic to temperate and montane-tropical. Genera: *Baeomyces* (Plate 2E), *Cladonia* (Fig. 1.10), *Pilophorus* (Fig. 1.8a), *Pycnothelia. Cladonia* is probably the most commonly collected lichen.

Umbilicariaceae.[158] Thallus foliose, umbilicate; apothecia lecideine. Habitat: rocks. Distribution: arctic to temperate. Genera: *Actinogyra, Agyrophora, Dermatiscum, Lasallia* (Fig. 1.6), *Omphalodiscus, Umbilicaria.*

Diploschistaceae. Thallus crustose; apothecia lecanorine, immersed to adnate. Habitat: rocks and soil. Distribution: arctic to temperate. Genera: *Conotrema, Diploschistes.*

Pertusariaceae. Thallus crustose; apothecia enclosed in thalline warts, spores large and thick-walled. Habitat: trees. Distribution: arctic to temperate and subtropical. Genera: *Melanaria, Perforaria, Pertusaria.*

Acarosporaceae. Thallus crustose to areolate-squamulose; apothecia immersed to adnate, spores minute, numerous. Habitat: rocks. Distribution: arctic to temperate. Genera: *Acarospora* (Plate 4B), *Biatorella, Glypholecia, Maronea, Sarcogyne, Sporastatia, Thelocarpon.*

Lecanoraceae. Thallus crustose, rarely squamulose or peltate; apothecia lecanorine. Habitat: trees and rocks. Distribution: arctic to temperate and montane-tropical. Genera: *Agrestia, Candelariella, Haematomma, Icmadophila, Ionaspis, Lecania, Lecanora* (Plate 9A), *Ochrolechia, Phlyctella, Phlyctidia, Phlyctis, Placopsis, Squamarina, Solenopsora.* This is the second largest lichen family, intergrading in many respects with the Lecideaceae.

Parmeliaceae. Thallus foliose to subfruticose, well developed; apothecia lecanorine, adnate to short-stalked. Habitat: trees and rocks. Distribution: arctic to temperate and tropical. Genera: *Anzia, Asahinea, Candelaria, Cavernularia, Cetraria* (Plate 16B), *Cetrelia, Hypogymnia, Menegazzia, Omphalodium, Parmelia* (Fig. 1.7; Plate 16C), *Parmeliopsis, Platismatia, Pseudevernia* (Plate 16D). *Parmelia* is the most commonly collected and widespread foliose lichen genus.

Usneaceae. Thallus fruticose, tufted or pendulous; apothecia lecanorine, spores simple (2-celled only in *Ramalina*), colourless. Habitat: trees and rocks. Distribution: arctic to temperate and montane-tropical. Genera: *Alectoria, Cornicularia, Dactylina, Evernia, Letharia, Neuropogon, Ramalina* (Fig. 1.8d, e), *Siphula, Thamnolia* (Fig. 1.8b), *Usnea* (Plate 5A, D, E). *Usnea*, often called Old Man's Beard, is probably the commonest fruticose lichen genus.

Physciaceae.[162, 264] Thallus crustose or foliose; apothecia lecanorine or lecideine, spores brown, two-celled or polarilocular. Habitat: trees and rocks. Distribution: boreal to tropical. Genera: *Anaptychia, Buellia, Physcia* (Plate 1B), *Pyxine, Rinodina* (Plate 4D), *Tornabenia.*

Teloschistaceae.[16] Thallus crustose, foliose or fruticose; apothecia lecanorine, disc usually orange (anthraquinones present), spores colourless, polarilocular. Habitat: trees and rocks. Distribution: arctic to temperate. Genera: *Caloplaca, Blastenia, Fulgensia, Gasparrinia, Teloschistes, Xanthoria.*

Fruiting bodies perithecia:

Order Sphaeriales

Pyrenulaceae.[232] Thallus crustose; phycobiont *Trentepohlia*; perithecia solitary; spores septate, the cells usually lens-shaped. Habitat: trees or rocks. Distribution: tropical to subtropical. Genera: *Belonia, Campylothelium, Microtheliopsis, Pleurotrema, Pyrenula* (Plate 10C), *Thelopsis.* Most of the genera formerly assigned to this little known family have been found to be ascolocular and provisionally transferred to the family Arthopyreniaceae.[223]

Strigulaceae.[232] Thallus crustose; spores septate, the cells cubical. Habitat: leaves of trees or bark and rocks. Distribution: tropical. Genera: *Clathroporina, Porina, Strigula.*

Verrucariaceae.[260] Thallus crustose, squamulose, or rarely umbilicate; paraphyses usually gelatinizing or lacking at maturity. Habitat: rocks, trees and soil. Distribution: arctic to temperate. Genera: *Dermatocarpon, Endocarpon, Heterocarpon, Normandina, Polyblastia, Staurothele, Thelidium, Thrombium, Trimmatothele, Verrucaria.* This large difficult family includes many aquatic or marine species.

Fruiting bodies mazaedia:

Order Caliciales

Caliciaceae. Thallus crustose, poorly differentiated; mazaedia terminal on

stipes. Habitat: bark, wood and soil. Distribution: temperate. Genera: *Calicium, Chaenotheca* (Plate 10A, B), *Coniocybe, Stenocybe.*
Cypheliaceae. Thallus crustose; mazaedium adnate. Habitat: trees or wood. Distribution: temperate. Genera: *Cypheliopsis, Cyphelium, Pyrgillus.*
Sphaerophoraceae. Thallus fruticose; mazaedia terminal, often partially enclosed by a thalline receptacle. Habitat: trees and humus. Distribution: arctic to temperate and montane-tropical, reaching highest development in the southern hemisphere. Genus: *Sphaerophorus.*

Asci with two walls (bitunicate), regularly or irregularly arranged in an ascostroma (pseudothecium) with branched pseudoparaphyses:

SUBCLASS LOCULOASCOMYCETIDAE

Pseudothecia poorly differentiated, asci irregularly distributed:

Order Myrangiales

Arthoniaceae. Thallus crustose, undifferentiated; pseudothecia lacking or irregular in outline. Habitat: trees. Distribution: temperate to tropical. Genera: *Arthonia, Arthothelium.*
Myrangiaceae. Thallus crustose, undifferentiated; pseudothecia resembling perithecia, grouped, opening through irregular pores. Habitat: trees. Distribution: temperate to tropical. Genera: *Dermatina, Mycoporellum.*

Pseudothecia well delimited, resembling perithecia, asci more or less regularly arranged in the stromatic layer:

Order Pleosporales

Arthopyreniaceae. Thallus crustose, poorly differentiated; pseudothecia resembling perithecia, solitary or clustered. Habitat: trees and rocks. Distribution: subboreal to tropical. Genera: *Anthracothecium, Arthopyrenia, Laurera, Leptorhaphis* (Plate 2D), *Melanotheca* (Plate 10D), *Microglaena, Microthelia, Parmentaria, Polyblastiopsis, Pseudopyrenula, Tomasellia, Trypethelium.*

Pseudothecia well delimited, round and resembling apothecia, lirelliform, or irregular in outline:

Order Hysteriales

Lecanactidaceae. Thallus crustose, poorly differentiated; pseudothecia round to elongate, with a thalline margin. Habitat: trees. Distribution: temperate to tropical. Genera: *Lecanactis, Platygraphopsis, Schismatomma.*
Opegraphaceae.[232] Thallus crustose; pseudothecia immersed to adnate, lirelliform, with an excipular margin. Habitat: trees or rocks. Distribution: temperate to tropical. Genera: *Chiodecton, Enterographa, Helminthocarpon, Mazosia, Opegrapha* (Plate 9D), *Sclerophyton.*
Roccellaceae.[280] Thallus crustose or fruticose; pseudothecia immersed to adnate, with a thalline exciple. Habitat: rocks and trees. Distribution: temperate to subtropical. Genera: *Dendrographa, Dirina, Reinkella, Roccella* (Fig. 1.8c), *Schizopelte.* The fruticose *Roccellas* are restricted to the dry coastlines of the eastern Pacific, African and Mediterranean regions.

Fruiting bodies containing basidia:

CLASS BASIDIOMYCETES[267]

Herpothallaceae. Thallus crustose; phycobiont *Trentepohlia*; fruiting bodies unknown. Habitat: trees. Distribution: tropical America. Genus: *Herpothallon*.

Coraceae. Thallus (fruiting body) smooth, bracket-shaped (Fig. 2.9a); phycobiont *Chlorococcum* or *Scytonema*. Habitat: rocks and soil. Distribution: tropical America and Asia. Genus: *Cora*.

Dictyonemataceae. Thallus (fruiting body) bracket-shaped, membranous, hirsute; phycobiont *Scytonema*. Habitat: trees and rocks. Distribution: tropical America and Asia. Genus: *Dictyonema*.

Clavariaceae.[216] Thallus *Clavaria*-like, dactyloid. Habitat: soil. Distribution: temperate Europe. Genera: *Botrydina, Clavulinopsis*.

Tricholomataceae. Thallus squamulose. Habitat: humus. Distribution: arctic to boreal. Genus: *Omphalina*.

Fruiting bodies unknown; thallus crustose to squamulose,
poorly differentiated:

CLASS FUNGI IMPERFECTI

Habitat: soil, rocks and trees. Distribution: arctic to temperate or tropical. Genera: *Cystocoleus, Lepraria* (Plate 1A), *Lichenothrix, Racodium*.

TAXONOMIC DELIMITATION OF LICHENS

Botanists rarely fail to distinguish correctly between lichens and fungi. The more highly structured, mineral grey or colourful lichen thallus bears little resemblance to the wispy often evanescent mycelium of true fungi. There are, however, organisms that are claimed by both mycologists and lichenologists. Most of these are crustose species which are not closely associated with algae. *Dermatina*, for example, is accepted as a lichen genus, but mycologists prefer to assign it to *Cyrtidula*, a non-lichenized genus.[80] Similarly *Placographa flexella* is classified as a fungus, but lichenologists recognize it as *Lithographa flexella*. Another group of interesting borderline lichens are parasitic on other fungi, as *Calicium polyporeum* on species of *Polyporus*, identical to fully lichenized forms of the genus, but not associated with symbiotic algae. Examples such as these show how easily the gap between lichenized and non-lichenized fungi can be bridged, but they fortunately account for only a minor proportion of all lichens.

Family characters

If separation of lichens and fungi is sometimes difficult at the generic level, it is no less so at the family level. As a general rule, both mycologists and lichenologists are inclined to recognize lichenization as a family character and maintain parallel families, as Pyrenulaceae (lichenized) and

Sphaeriaceae (non-lichenized).[151] Yet it is interesting to note that mycologists include within the order Lecanorales both lichenized and non-lichenized fungi.[80] This includes not only free-living saprophytes such as *Agyrium* and *Pseudographis* but also the parasymbiotic lichen fungi, of which *Abrothallus* (Plate 9F) and *Karschia* are well-known examples. Other parasymbionts are classified under appropriate fungal families, as *Discothecium*, in the family Didymosphaeriaceae and *Pharcidia* in the Pleosporaceae.

There seems to be little doubt that families should be characterized by the ontogeny and structure of the ascocarps. Henssen[140] has laid the groundwork for a new alignment of genera in the family Peltigeraceae by using hemiangiocarpous apothecial development as a primary character. Lamb[165] has segregated *Stereocaulon* from the family Cladoniaceae as a new family, Stereocaulaceae, since the fruticose pseudopodetia, although superficially the same as the podetia of *Cladonia*, have a thalline origin. Modifications of the ascocarp by lichenization provide an additional basis for separating families and even genera. On such a basis, groups with lecanorine and lecideine apothecia are split off into different families, even when no spore or hymenial differences exist.

Many lichen families date from the mid-nineteenth century when growth form was considered an important character. Crustose genera were placed in one family and non-crustose genera in another. The Buelliaceae and Physciaceae, the Caloplacaceae and Teloschistaceae, the Verrucariaceae and Dermatocarpaceae, and the Lecanoraceae and Parmeliaceae, to mention but a few, are parallel crustose-non-crustose families, even though within each pair there are no differences in ascocarp structure. Recent workers have tended to assign less importance to growth form and now recognize single families, the Physciaceae, the Teloschistaceae[16] and the Verrucariaceae. The family Lecanoraceae is still maintained separately from the foliose Parmeliaceae. Renunciation of growth form, if rigorously applied, would further reduce the number of lichen families, and we can expect that taxonomists will expend a considerable amount of effort in this direction in the near future.

Generic characters

The generic limits of lichens as a group, when compared with non-lichenized fungi, are probably too conservative. Approximately 15 000 species of lichens are accommodated by only 400 genera while nearly the same number of fungi are distributed among 1820 genera.[14] Recent workers are beginning to split up larger heterogeneous genera, as removing the lobate *Lecanora* species to the genus *Squamarina*. Squamulose *Lecideas* are frequently placed in the genus *Psora*. The generic limits of *Parmelia*, the commonest foliose genus with over 600 species, have been extremely broad, but there is a tendency now to accept the generic segregates *Hypogymnia*, *Pseudevernia* and *Cavernularia*.

The most useful, though not necessarily most fundamental, generic character lies in the ascospores. Spore features such as colour and septation are relatively constant and, as in non-lichenized fungi, form a conservative basis for classification. There are of course intergradations where transversely septate spores have one or two longitudinal septa and fall within the range of genera with muriform spores. This is especially troublesome in the Graphidaceae[284] where generic separation according to spores appears to be artificial.

Thallus growth form is frequently used as a taxonomic character and, although it does not seem particularly sound, there is little chance that it can wholly be rejected at the generic level. Examples of generic splitting by growth form can be cited in most lichen families. Within the Teloschistaceae, for instance, four genera may be recognized: *Caloplaca* (crustose), *Gasparrinia* (crustose-lobate), *Xanthoria* (foliose) and *Teloschistes* (fruticose). The ascocarps are more or less identical and, if we can believe Tomaselli's physiological studies, *Gasparrinia* and *Xanthoria* may be manifestations of a single fungus.[268] Algae-based foliose genera recognized in the nineteenth century have now been discredited. The distinctions between *Sticta* (green phycobiont) and *Stictina* (blue-green), *Lobaria* and *Lobarina*, *Peltigera* and *Peltidea*, or *Nephroma* and *Nephromium*, are no longer seriously upheld.

Species characters

When we speak of a 'species' of lichen, we are referring only to the fungal component in either a lichenized or unlichenized state, according to Article 13 of the 1961 *International Code of Botanical Nomenclature*. By this definition, a separate name for the isolated mycobiont, for example, *Xanthoriomyces parietinae* isolated from *Xanthoria parietina*, is superfluous.

As to the delimitation of species, there is no unanimity of opinion. Species are recognized by more or less arbitrary combinations of vegetative and ascocarp characters. The more complex foliose and fruticose species are usually separated by the presence or absence of soredia, isidia, cilia, pseudocyphellae and lobules, type of rhizines and other thalline characters. Spores vary but little and play a relatively small role. Crustose species, on the other hand, lack a variety of really distinctive thalline characters, and the structure of the exciple, spore size and septation, and other microscopical traits are paramount. In both groups chemical features are being used to an increasing extent in taxonomy.

The number of lichen species is usually calculated from Zahlbruckner's catalogue, which lists all lichen names published to about 1930.[286] It contains approximately 15 000 species. Lamb's supplement added another

5684 names for a total of some 20 000 species.[169] There is good reason to suspect that this total is inflated since monographic studies in lichens are not as advanced as in other plant groups. Recent work has uncovered a high degree of synonymy. In *Parmelia* subgenus *Amphigymnia*, for example, about 70% of the 350 published names are synonyms or otherwise invalid.[124] Only 106 species are now recognized. Nearly half of the 158 species names in *Stereocaulon* are synonyms.[171] The degree of synonymy in the large difficult crustose genera can only be imagined. Applying a conservative estimate that 25% of the published names are synonyms, we are probably correct in saying that there are about 15 000 distinct species of lichens and that the number of undiscovered species yet to be described is only a small fraction of this total.

IDENTIFICATION OF LICHENS

Lichens are basically no more difficult to study than any other group of cryptogams. Many of the characters that separate species are qualitative and decisions often require no more than a hand-lens. Specimens may be collected at any time of the year and little care is needed for preparation. The greatest obstacle, however, is the lack of useful illustrated manuals for field and laboratory identification. There are a number of lichen Floras for the European countries but only a few are intended for beginners. Lichens neither reproduce well in photographs nor lend themselves to simple line drawings; written descriptions and keys alone are highly unsatisfactory.

Duncan's *A Guide to the Study of Lichens*[82] and the more complete *Introduction to British Lichens* by Duncan and James[82A] are recommended to British students. The British Lichen Society issues a journal, *The Lichenologist*, which contains more technical articles on lichens for advanced students. In America, Hale's *How to Know the Lichens*[127B] covers the foliose and fruticose species. Keys to crustose lichens may be found in Fink's *Lichen Flora of the United States*[92] and in Hale's *Lichen Handbook*.[120] Articles on lichenology in general and lists of current literature compiled by W. L. Culberson appear in each issue of the *Bryologist*, published by the American Bryological and Lichenological Society. A number of floras and manuals in French or German may be consulted for European lichens.

II

Economic Uses and Applications

LICHENS AS FOOD

Lichens have not been utilized as a food source by man to any great extent.[184] The thalli, though edible, are often tasteless and may contain bitter irritating acids, especially fumarprotocetraric acid, that must first be leached out by boiling in soda. Their food value, however, compares favourably with that of cereal crops. In times of famine peoples in boreal or subarctic regions have resorted to lichens as a supplemental source of carbohydrates, mixing them with flour or boiling to extract gelatins. In Japan the foliose rock tripes (*Umbilicaria*), called Iwatake, are eaten in salads or fried in deep fat; they are considered a delicacy.

Lichens are important as food for animals in the arctic regions. Llano[184] reports deer nibbling *Usneas* on conifers in boreal forests and lemmings consuming tundra lichens. Reindeer and caribou supplement their normal diet of sedges and willow twigs with lichens during the winter. They forage on fruticose soil lichens, the *Cladonias* and *Cetrarias*, and, if the snow is too deep, on corticolous species above the snow level.[84] Scotter[242] has analysed some forage lichens and found them to be low in protein, calcium and phosphorus in relation to the estimated requirements of caribou in Canada. Nevertheless, lichens may constitute as much as 95% of the total diet under severe conditions but probably normally amount to less than two-thirds of the total. Laplanders practise controlled grazing and even harvesting of reindeer mosses in northern Scandinavia. Vast areas in subarctic Canada are covered with lichen fields and have a considerable potential for use as reindeer pastures.

Sheep in the Libyan deserts are reported to graze on the subfoliose lichen

Lecanora esculenta. This lichen forms a thick loose crust on soil and rocks in surprising abundance and is easily eaten by the sheep, although they eventually suffer premature loss of teeth by abrasion.[120] This same lichen is suspected of being the fabled manna of the ancient Israelites.[131]

Lichens appear to be a common source of food for land snails and slugs. One does not have to look long to find lichen thalli with tell-tale rasp marks of snails that have chewed away the upper cortex or parts of lobes. Plitt[215] made an interesting study of the common tropical snail *Oxystyla undata* and showed that it can live on the thalli of common temperate lichens. Psocids, lice-like insects that infest larch trees, also eat lichens and under controlled laboratory conditions will survive on *L. conizaeoides*.[49] Tardigrades are regularly associated with and probably consume lichens on trees.[74] A number of invertebrates secrete lichenase, an enzyme that rapidly breaks lichenin down into glucose,[197] and this would explain their ability to utilize lichens as food so readily.

MEDICAL ASPECTS

During the Middle Ages lichens were held in high regard by medical practitioners. *Lobaria pulmonaria* was used to treat lung diseases because of its superficial resemblance to lung tissue; *Parmelia sulcata* was eagerly sought as a remedy for cranial maladies. A widely used prescription for rabies called for a half-ounce of powdered *Peltigera* mixed with two drachmas of black pepper. This mixture was taken on four consecutive days in a half-pint of warm milk. When Linnaeus named this particular lichen in 1753, he called it *Lichen caninus*, the dog lichen.[120]

The use of lichens in folk medicine has persisted into modern times. The Seminole Indians in Florida and the Chinese employ various lichens in medicine, especially as expectorants. *Usnea* species are most frequently used. Ahmadjian and Nilsson[11] report that the Iceland moss *Cetraria islandica* is widely sold in apothecary shops in Sweden and is claimed to be effective in treating diabetes, lung diseases and catarrh. *Peltigera canina* is eaten in India as a remedy for liver ailments[259] and its high content of the amino acid methionine may contribute to its alleged curative power.

Antibiotic properties

As with other plants used in folk medicine, some lichens do contain active principles which can explain their efficacy. In 1944 Burkholder *et al.*[56] included lichens in a general programme of assaying plants for new antibiotics and discovered that extracts from 52 different species in eastern North America inhibited growth of several kinds of bacteria. This led to a feverish race to discover antibiotics among the lichens, and in the 10-year period until 1954 practically all available lichen species and substances were tested. The inhibitory agent has almost invariably been identified as one of the lichen substances. As a general rule, they are ineffective against gram-negative bacteria (*Escherichia*, *Salmonella* and *Shigella*).[257] Gram-positive bacteria are more or less strongly inhibited by usnic acid, protolichesterinic acid and a few orcinol derivatives; β-orcinol derivatives and the pulvic acid group are relatively inactive. Unidentified biologically active compounds have also been isolated from mycobiont cultures.[12]

The most promising lichen substance is the yellow pigment usnic acid, a broad spectrum antibiotic. It strongly inhibits *Mycobacterium*[257] and has found commercial use as the active ingredient in a salve called 'Usno'. This medication, now widely available in Europe, is more effective than penicillin salves in treating external burns and superficial wounds. Similar commercial preparations are being marketed in Germany as 'Usniplant' and in Russia as 'Binan'. The sources of usnic acid for these drugs are the yellow species of *Cladonia*, reindeer mosses that grow abundantly in Scandinavia and northern Europe.

Lichen substances are also being investigated experimentally as anti-biotics in plant pathology. Sodium usnate, for example, has been found effective against the tomato canker (*Corynebacterium michiganensis*).[20] Vulpinic, physodic, salacinic and usnic acids all show a moderate degree of inhibition of the blue-staining wood fungus *Trichosporium*.[34] Tobacco mosaic virus is inhibited by various lichen extracts and proven active principles are lecanoric, psoromic and usnic acids.[96]

The antifungal properties of lichen substances are not well known and much work remains to be done in this field. It has been discovered recently that the growth of the mould *Neurospora crassa* is strongly inhibited by usnic acid, and haematommic acid, a monocyclic phenol derivative of lichen depsides, is also effective.[143] Salacinic and evernic acids have little or no antifungal activity. This may provide an explanation of why im-properly dried herbarium specimens of *Parmelia saxatilis* or *P. sulcata*, species containing salacinic acid, are especially susceptible to moulding. With the same treatment, species which contain usnic acid accompanied by salacinic acid, such as some of the yellow *Parmelias* and *Ramalinas*, are far more resistant to infestation.

Harmful effects

One of the more unusual economic effects of lichens has recently been brought to light by dermatologists.[196] Woodcutters in Canada have long been susceptible to a skin rash, and the allergen has recently been traced to usnic acid, the yellow pigment of the common corticolous *Usneas* and *Evernias*. Dust of soredia on clothing has even caused allergic reactions in wives of woodcutters not directly exposed in the woods. The possibilities that usnic acid may be a photosensitizer and a cause of respiratory allergy are also being explored.

Arctic lichens are having a much more deleterious effect on Eskimos and Laplanders. Reindeer mosses at high latitudes have accumulated large quantities of radioactive caesium and strontium from the fall-out of atom-bomb tests. The animals eat these lichens and soon build up a high body burden of radioactivity. Since the arctic peoples live on caribou and rein-deer, the radioactivity is passed along to them in the food chain. Eskimos in Alaska have body burdens as high as one-third of the maximum per-missible safe level,[210] and Laplanders have accumulated comparably high levels. Although no symptoms of radiation illness have been seen and bomb testing has fortunately ceased, public health authorities are deeply con-cerned about possible after-effects.

LICHENS AS DYESTUFFS

Before the discovery of coal-tar dyes, lichens had considerable economic importance as dyestuffs. They were mentioned by the ancient Greeks and apparently used widely in the Mediterranean region, especially as the source of a valuable purple dye. From the early thirteenth century to 1800 there was an active dye industry in southern Europe where the fruticose lichen *Roccella* (Fig. 1.8c) was extensively collected as a raw material. In northern Europe the locally abundant species of *Parmelia*, *Evernia* and *Ochrolechia* (crottal) were collected and even to this day support a small cottage-type dyeing industry in Scandinavia.

Bolton[46] has published a very useful account on lichen dyes that should be consulted by those interested in preparing dyestuffs. The lichen thalli are pulverized, boiled in water and treated with ammonia, which in olden times was derived from urine. The basic dyeing principle can be extracted as a paste, orchil, that colours wool many delicate shades of purple, red or brown. Harris tweeds manufactured in Scotland are still dyed with native lichen dyestuffs that impart a unique musty odour.

The chemistry of the mechanism of lichen dyes has only recently been worked out. The basic depsides that make up the raw ingredients of orchil paste are split into monocyclic orcin and, following ammonification, the dyestuff orcein. Orcein is a mixture of three compounds, 7-oxyphenoxazon (Fig. 11.1), 7-aminophenoxazon and 7-aminophenoxazin.[203] Solberg[255] discovered that the tawny yellow-brown to reddish colours produced on wool and silk by lichen dyestuffs are traceable to hydroxyaldehydes. The aldehyde radical reacts with the free amino acids of wool and forms stable azomethine linkages.

The amphoteric dye litmus, a familiar acid-base indicator in chemistry laboratories, is derived from depside-containing lichens. It is related to orchil but represents a more complex mixture of polymeric compounds with the 7-oxyphenoxazon chromophore and its anion, as shown in Fig. 11.1.[203]

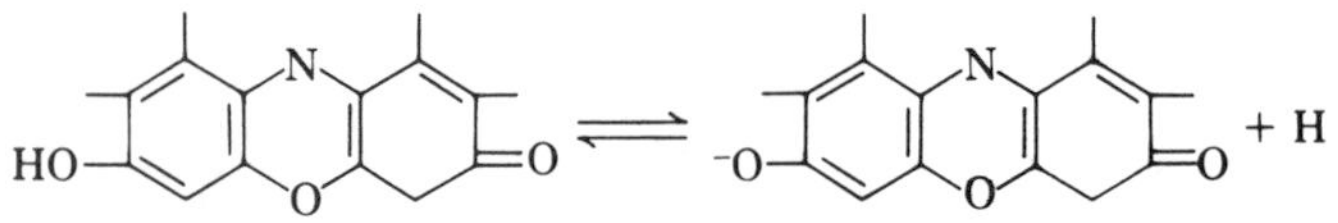

Fig. 11.1   Structural formula of 7-oxyphenoxazon and its anion. (After Musso[203])

EFFECTS ON THE SUBSTRATUM

Lichens, though autotrophic, are often in such intimate contact with their substrates that it might be supposed that they could have harmful effects. In careful studies on the attachment organs of corticolous species of *Parmelia* and *Ramalina*, Porter[217] was able to find extensive penetration of the rhizines through the cork, cortex, bast and cambium, as far as the living wood. The lichen hyphae blocked lenticels, split the cork layers horizontally, and by increasing air exchange in the bark indirectly caused the cork cells to thicken and become more permeable to water. Small shrubs and trees densely covered with lichens could clearly be stunted and damaged.

It is common practice in some orchards in Europe and southern United States to destroy corticolous lichen growths on fruit trees with fungicides in the belief that lichen-free trees are more robust. Even if direct damage by rhizines is of no consequence, a dense cover of lichens can harbour and provide shelter for many harmful insects and arachnid pests. Foliicolous lichens have not been proved to parasitize leaves[232] but their very abundance in tropical regions on the leaves of commercially important trees has been a source of concern to plant pathologists.

An interesting destructive effect of lichens on pane glass has been reviewed by Mellor.[193] In France and England the windows of churches dating back to the thirteenth century have been seriously corroded by lichen growth. Mellor concludes that mechanical action is primarily responsible for this damage, although chemical action too is implicated. The drier climate of North America has apparently prevented similar growth on pane glass.

Saxicolous lichens are suspected to have a comparable role in the breakdown of rocks, although their ability to form soil is probably exaggerated. Crustose lichen thalli are often interspersed with microscopic rock fragments which have been loosened by mechanical contraction and expansion when the thallus is alternately moistened and dried.[99] The substances excreted by the thalli are obviously too weak to alter the rocks by hydrogen ion exchange, but chelation could be an important mechanism in mineral breakdown (Fig. 11.2).[235] On the basis of molecular structure lichen substances appear to be excellent chelators and several studies have been directed at their possible commercial application.

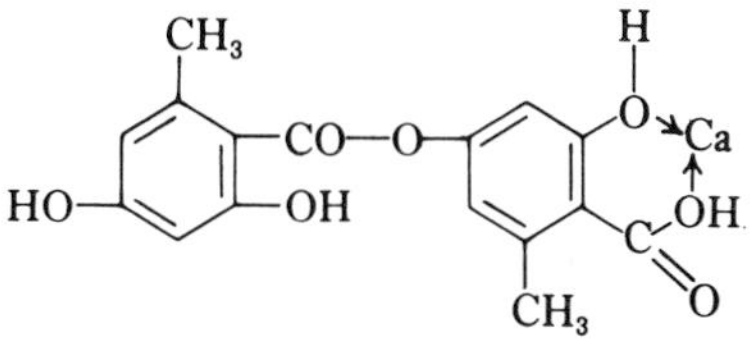

Fig. 11.2 Theoretical mechanism for the chelation of calcium by lecanoric acid.

MISCELLANEOUS USES

Various essential oils and derivatives of depsides are extracted from species of *Evernia*, *Parmelia* and *Ramalina*. Some have agreeable odours and are in demand for scenting soaps and for an essence in better perfumes. The commercial collection and extraction of these 'oak mosses' for the perfume industry are still actively carried on in southern France and Yugoslavia.[195]

Lichens have found some practical use as indicators in geological exploration. The occurrence of bright lemon-yellow *Cetrarias*, especially *C. tilesii*, is highly correlated with marble and limestone deposits. Spectrographic analyses of saxicolous lichens in Colorado[177] have shown that the thalli contain high concentrations of rare mineral elements which could lead to the discovery of ore bodies rich in these elements.

One of the bizarre uses to which lichens have been put is as packing materials for mummies in ancient Egypt.[40] In the southern Sahara, corticolous lichens are used in pipe mixtures for smoking, in particular *Parmelia andina*, which is known to contain large amounts of lecanoric acid.[174]

Fruticose *Cladonias* are frequently used as facsimile trees and shrubbery for architectural models and model trains. They are dyed green and treated with glycerine so as to remain soft and pliable. Florists often add *Cladonias* to floral arrangements and miniature gardens for decorative effect.

References

1. ABBAYES, H. DES (1951). *Traité de Lichénologie*, Lechevalier, Paris.
2. AHMADJIAN, V. (1958). *Bot. Notiser*, **111**, 632.
3. AHMADJIAN, V. (1959), *Svensk bot. Tidskr.*, **53**, 71.
4. AHMADJIAN, V. (1960). *Bryologist*, **63**, 250.
5. AHMADJIAN, V. (1962). *Am. J. Bot.*, **49**, 277.
6. AHMADJIAN, V. (1963). *Scient. Am.*, **208**, 122.
7. AHMADJIAN, V. (1964). *Bryologist*, **67**, 87.
8. AHMADJIAN, V. (1965). *A. Rev. Microbiol.*, **19**, 1.
9. AHMADJIAN, V. (1966). *Science, N.Y.*, **151**, 199.
10. AHMADJIAN, V. (1966). Unpublished data.
10A. AHMADJIAN, V. (1967). *The Lichen Symbiosis*, Blaisdell, Waltham.
10B. AHMADJIAN, V. and HEIKKILÄ, H. (1970). *Lichenologist*, **4**, 259.
11. AHMADJIAN, V. and NILSSON, S. (1963). Swedish Lichens. *Yb. Am. Swed. Hist. Fdn.*
12. AHMADJIAN, V. and REYNOLDS, J. T. (1961). *Science, N.Y.*, **133**, 700.
13. AHTI, T. (1959). *Suomal. eläin- ja kasvit. Seur. van. Tiedon.*, **30** (4), 1.
14. AINSWORTH, G. C. (1961). *Dictionary of the Fungi*, Commonwealth Mycol. Inst., Kew, Surrey.
15. ÅKERMARK, B. (1961). *Acta chem. scand.*, **15**, 1695.
16. ALMBORN, O. (1963). *Bot. Notiser*, **116**, 161.
17. ALTMAN, P. L. and DITTMER, D. S. (1964). *Biological Data Book*, Federation of the American Society for Experimental Biology, Washington.
18. ALVIN, K. L. (1960). *J. Ecol.*, **48**, 331.
19. ANDERSON, E. and RUDOLPH, E. D. (1956). *Evolution, Lancaster, Pa.*, **10**, 147.
20. ARK, P. A., BOTTINI, A. T. and THOMPSON, J. P. (1960). *Pl. Dis. Reptr*, **44**, 200.
21. ASAHINA, Y. (1936). *J. Jap. Bot.*, **12**, 516. This is the first of a series of articles on identification that appeared regularly in this journal until 1940.
22. ASAHINA, Y. (1936). *J. Jap. Bot.*, **12**, 690.
23. ASAHINA, Y. (1937). *J. Jap. Bot.*, **13**, 219.
24. ASAHINA, Y. (1937). *Bot. Mag., Tokyo*, **41**, 759.
25. ASAHINA, Y. (1939). *Illustrated Flora of Japanese Cryptogams*, Sanseido, Tokyo.
26. ASAHINA, Y. (1939). *J. Jap. Bot.*, **15**, 205.
27. ASAHINA, Y. (1939). *J. Jap. Bot.*, **15**, 353.

28. ASAHINA, Y. (1943). *J. Jap. Bot.*, **19**, 227.
29. ASAHINA, Y. (1952). *J. Jap. Bot.*, **27**, 239.
30. ASAHINA, Y. (1956). *Lichens of Japan III. Usnea.* Research Institute for Natural Resources, Tokyo.
31. ASAHINA, Y. (1963). *J. Jap. Bot.*, **38**, 255.
32. ASAHINA, Y. (1964). *J. Jap. Bot.*, **39**, 209.
33. ASAHINA, Y. and SHIBATA, S. (1954). *Chemistry of Lichen Substances*, Japan Society for the Promotion of Science, Tokyo.
34. ASCORBE, F. J. (1953). *Caribb. Forester*, **14**, 136.
35. ASPINALL, G. O., HIRST, E. L. and WARBURTON, M. (1955). *J. chem. Soc.*, **1955**, 651.
36. BACHMANN, O. (1963). *Öst. bot. Z.*, **110**, 103.
36A. BAILEY, R. H. (1970). *Lichenologist*, **4**, 256.
37. BARKMAN, J. (1958). *On the Ecology of Cryptogamic Epiphytes*, Van Gorcum & Co., The Hague.
38. BARNETT, H. L. (1964). *Mycologia*, **56**, 1.
39. BARY, A. DE (1866). *Morphologie und Physiologie der Pilze, Flechten und Myxomyceten*, Leipzig.
40. BAUMANN, B. B. (1960). *Econ. Bot.*, **14**, 84.
41. BECKMANN, P. (1907). *Engl. Bot. Jb. Beibl.*, **88**, 1.
42. BEDNAR, T. W. and HOLM-HANSEN, O. (1964). *Pl. Cell Physiol.*, Tokyo, **5**, 297.
43. BESCHEL, R. E. (1958). *Ber. naturw.-med. Ver. Innsbruck*, **52**, 1.
44. BESCHEL, R. E. (1961). In *Geology of the Arctic*, University of Toronto Press, Toronto.
45. BESCHEL, R. E. (1965). Unpublished data.
46. BOLTON, E. M. (1960). *Lichens for Vegetable Dyeing*, Charles T. Branford Co., Newton Centre, Mass.; Studio Vista, London.
47. BONNIER, G. (1886). *C.r. hebd. Séanc. Acad. Sci.*, Paris, **103**, 942.
48. BRIGHTMAN, F. H. (1959). *Lichenologist*, **1**, 104.
49. BROADHEAD, E. (1958). *J. Anim. Ecol.*, **27**, 217.
50. BRODO, I. M. (1961). *Am. Midl. Nat.*, **65**, 290.
51. BRODO, I. M. (1961). *Ecology*, **42**, 838.
52. BRODO, I. M. (1964). *Bryologist*, **67**, 76.
53. BRODO, I. M. (1965). *Bryologist*, **68**, 451.
54. BRODO, I. M. (1966). Unpublished data.
55. BROWN, C. J., OLLIS, W. D. and VEAL, P. L. (1960). *Proc. chem. Soc.*, **1960**, 393.
56. BURKHOLDER, P. R., EVANS, A. W., MCVEIGH, I. and THORNTON, H. K. (1944). *Proc. natn. Acad. Sci. U.S.A.*, **30**, 250.
57. BURLEY, J. W. A. B., GILBERT, G. E. and CLUM, C. C. (1962). *Spec. Bull. Ohio agric. Exp. Stn*, **10**, 1.
58. BURNETT, J. H. (1956). *New Phytol.*, **55**, 50.
59. CASTLE, H. and KUBSCH, F. (1949). *Archs Biochem.*, **23**, 158.
60. ČERNOHORSKÝ, Z., NÁDVORNÍK, J. and SERVÍT, M. (1956). *Klíč K Určovani Lišejniků CSR*, Československé Akademie Věd, Prague.
61. CHADEFAUD, M. (1940). *Revue Mycol.*, **5**, 87.
62. CLEMENTS, E. and SHEAR, L. (1957). *The Genera of Fungi*, Hafner, London & New York.
62A. COKER, P. D. (1967). *Lichenologist*, **3**, 428.
63. CULBERSON, C. F. (1963). *Microchem. J.*, **7**, 153.
64. CULBERSON, C. F. (1965). *Bryologist*, **68**, 435.
65. CULBERSON, C. F. (1965). *Phytochem.*, **4**, 951.

65A. CULBERSON, C. F. (1969). *Chemical and Botanical Guide to Lichen Products*, University of North Carolina Press, Chapel Hill.
65B. CULBERSON, C. F. (1972). *Bryologist*, **75**, 54.
66. CULBERSON, C. F. and CULBERSON, W. L. (1958). *Lloydia*, **21**, 189.
66A. CULBERSON, C. F. and KRISTINSSON, H. (1970). *J. Chromat.*, **46**, 85.
67. CULBERSON, W. L. (1955). *Ecol. Monogr.*, **25**, 215.
68. CULBERSON, W. L. (1958). *Phyton, Horn*, **11**, 85.
69. CULBERSON, W. L. (1961). *Taxon*, **10**, 69.
70. CULBERSON, W. L. (1962). *Nova Hedwigia*, **4**, 563.
71. CULBERSON, W. L. (1963). *Science, N.Y.*, **139**, 40.
72. CULBERSON, W. L. (1963). *Bryologist*, **66**, 224.
73. CULBERSON, W. L. and CULBERSON, C. F. (1956). *Am. J. Bot.*, **43**, 678.
73A. CULBERSON, W. L. and CULBERSON, C. F. (1967). *Science, N.Y.*, **158**, 1195.
73B. CULBERSON, W. L. and CULBERSON, C. F. (1968). *Contr. U.S. natn. Herb.*, **34**, 449.
73C. CULBERSON, W. L. and CULBERSON, C. F. (1970). *Bryologist*, **73**, 1.
74. CURTIN, C. B. (1957). *Proc. Pa. Acad. Sci.*, **31**, 142.
75. DAHL, E. (1952). *Revue bryol. lichen.*, **21**, 119.
76. DEGELIUS, G. (1935). *Acta phytogeogr. suec.*, **7**, 1.
77. DEGELIUS, G. (1945). *Svensk bot. Tidskr.*, **39**, 419.
78. DEGELIUS, G. (1954). *Symb. bot. upsal.*, **13** (2), 1.
79. DEGELIUS, G. (1964). *Acta Horti gothoburg.*, **27**, 11.
80. DENNIS, R. W. G. (1960). *British Cup Fungi and Their Allies*, Ray Society, London.
80A. DE SLOOVER, J. and LEBLANC, F. (1968). In *Proceedings of the Symposium on Recent Advances in Tropical Ecology*, Varanasi.
80B. DIBBEN, M. J. (1971). *Lichenologist*, **5**, 1.
81. DUGHI, R. (1954). *Revue bryol. lichen.*, **23**, 300.
82. DUNCAN, U. K. (1959). *A Guide to the Study of Lichens*, Buncle & Co. Ltd., Arbroath, Scotland.
82A. DUNCAN, U. K. and HAMES, P. (1970). *Introduction to British Lichens*, Buncle & Co. Ltd., Arbroath, Scotland.
83. DUVIGNEAUD, P. (1939). *Bull. Ass. fr. Avanc. Sci.*, **63**, 585.
84. EDWARDS, R. Y., SOOS, J. and RITCEY, R. W. (1960). *Ecology*, **41**, 425.
85. ELENKIN, A. (1902). *Izv. imp. S.-Peterb. bot. Sada*, **2**, 65.
86. ELFVING, F. (1930). *Acta Soc. Sci. fenn. NS*, **1**, 1.
87. ERBISCH, F. H. (1964). *J. Cell Biol.*, **23**, 28A.
88. ERTL, L. (1951). *Planta*, **39**, 245.
89. EVANS, A. W. (1943). *Bull. Torrey bot. Club*, **70**, 139.
90. FAMINTZIN, A. (1907). *Biol. Zbl.*, **27**, 353.
90A. FERRY, B. W., BADDELEY, M. S. and HAWKSWORTH, D. L. (1973). *Air Pollution and Lichens*, Athlone Press, London.
91. FINK, B. (1911). *Mycologia*, **3**, 231.
92. FINK, B. (1935). *The Lichen Flora of the United States*, The University of Michigan Press, Ann Arbor.
93. FISCHER, E. and FISCHER, H. O. L. (1913). *Ber. dt. chem. Ges.*, **46**, 1138.
94. FOLLMANN, G. (1960). *Flechtenleben*, R. Oldenbourg, Munich.
95. FOLLMANN, G. (1961). *Revta univ., Santiago*, **46**, 149.
96. FOLLMANN, G. and VILLAGRÁN, V. (1964). *Z. Naturw.*, **22**, 543.
96A. FOX, C. (1966). Unpublished data.
97. FREY, E. (1959). *Ergebn. wiss. Unters. schweiz. Natnparks*, **41**, 239.

98. FRITSCH, F. E. (1952). *Proc. R. Soc.*, ser. B, **139**, 185.

99. FRY, E. J. (1926). *Ann. Bot.*, **38**, 175.

99A. GALUN, M., BEN-SHAUL, Y. and PARAN, N. (1971). *New Phytol.*, **70**, 837.

100. GÄUMANN, E. A. (1952). *The Fungi*, translated by F. L. Wynd, Hafner, London and New York.

100A. GILBERT, O. L. (1971). *Lichenologist*, **5**, 26.

100B. GILBERT, O. L. (1971). *Lichenologist*, **5**, 11.

101. GILBERT, P. W. (1942). *Ecology*, **23**, 215.

102. GOEBEL, K. (1926). *Ber. dt. chem. Ges.*, **44**, 158.

103. GORHAM, E. (1959). *Can. J. Bot.*, **37**, 327.

104. GRASSI, M. M. (1950). *Lilloa*, **24**, 297.

105. GRESSITT, J. L., SEDLACEK, J. and SZENT-IVANY, J. J. H. (1965). *Science, N.Y.*, **150**, 1833.

106. GROENHART, P. (1965). *Persoonia*, **4**, 1.

107. GUSTAFSON, F. G. (1954). *Bull. Torrey bot. Club*, **81**, 313.

108. HALE, M. E. (1950). *Bryologist*, **53**, 181.

109. HALE, M. E. (1952). *Bull. Torrey bot. Club*, **79**, 251.

110. HALE, M. E. (1952). *Ecology*, **33**, 398.

111. HALE, M. E. (1955). *Ecology*, **36**, 45.

112. HALE, M. E. (1955). *Bryologist*, **58**, 242.

113. HALE, M. E. (1956). *Castanea*, **21**, 30.

114. HALE, M. E. (1956). *Bull. Torrey bot. Club*, **83**, 218.

115. HALE, M. E. (1956). *Am. J. Bot.*, **43**, 456.

116. HALE, M. E. (1956). *Science, N.Y.*, **123**, 671.

117. HALE, M. E. (1957). *Mycologia*, **39**, 417.

118. HALE, M. E. (1958). *Bull. Torrey bot. Club*, **85**, 182.

119. HALE, M. E. (1959). *Bull. Torrey bot. Club*, **86**, 126.

120. HALE, M. E. (1961). *Lichen Handbook*, Smithsonian Institution, Washington.

121. HALE, M. E. (1962). *Bryologist*, **65**, 291.

122. HALE, M. E. (1963). *Brittonia*, **15**, 126.

123. HALE, M. E. (1964). *Bryologist*, **67**, 462.

124. HALE, M. E. (1965). *Contr. U.S. natn. Herb.*, **36**, 193.

125. HALE, M. E. (1965). *Svensk bot. Tidskr.*, **59**, 37.

126. HALE, M. E. (1965). *Bryologist*, **68**, 193.

127. HALE, M. E. (1966). Unpublished data.

127A. HALE, M. E. (1966). *Israel J. Bot.*, **15**, 150.

127B. HALE, M. E. (1969). *How to Know the Lichens*, W. C. Brown Co., Dubuque.

127C. HALE, M. E. (1970). *Bryologist*, **73**, 72.

127D. HALE, M. E. (1972). *Proc. biol. Soc. Wash.*, **85**, 287.

127E. HALE, M. E. (1973). *Smithson. Contr. Bot.*, **10**, 1.

128. HALE, M. E. and CULBERSON, W. L. (1960). *Bryologist*, **63**, 137.

128A. HALE, M. E. and WIRTH, M. (1971). *Phytologia*, **22**, 36.

129. HARDER, R. and UEBELMESSER, E. (1958). *Arch. Mikrobiol.*, **31**, 82.

130. HARLEY, J. L. and SMITH, D. C. (1956). *Ann. Bot. N.S.*, **20**, 513.

131. HARRISON, S. G. (1950), *Kew Bull.*, **1950**, 407.

131A. HAWKSWORTH, D. L. (1969). *J. Microscopie*, **8**, 753.

131B. HAWKSWORTH, D. L. and CHAPMAN, D. S. (1971). *Lichenologist*, **5**, 51.

131C. HAWKSWORTH, D. L. and ROSE, F. (1970). *Nature, Lond.*, **227**, 145.

132. HAYNES, F. N. (1964). 'Lichens', in *Viewpoints in Biology*, **3**, CARTHY, J. D. and DUDDINGTON, C. L. Butterworths, London.

133. HENDRICKSON, J. R. and WEBER, W. A. (1964). *Science, N.Y.*, **144**, 1463.
134. HENRIKSSON, E. (1958). *Svensk bot. Tidskr.*, **52**, 391.
135. HENRIKSSON, E. (1960). *Physiologia Pl.*, **13**, 751.
136. HENRIKSSON, E. (1961). *Physiologia Pl.*, **14**, 813.
137. HENRIKSSON, E. (1963). *Physiologia Pl.*, **16**, 867.
138. HENRIKSSON, E. (1964). *Svensk bot. Tidskr.*, **58**, 361.
139. HENSSEN, A. (1963). *Symb. bot. upsal.*, **18**, 1.
140. HENSSEN, A. (1963). *Can. J. Bot.*, **41**, 1331.
141. HENSSEN, A. (1965). *Lichenologist*, **3**, 29.
141A. HENSSEN, A. (1970). *Dt. bot. Gesell. Neue Folge*, **4**, 5.
142. HERRE, A. W. (1904). *Bot. Gaz.*, **38**, 218.
143. HESS, D. (1959). *Z. Bot.*, **48**, 136.
144. HESS, D. (1959). *Z. Naturf.*, **14***b*, 345.
145. HILLMAN, J. (1920). *Annls mycol.*, **18**, 14.
146. HUNECK, S. (1963). *Naturwissenschaften*, **50**, 645.
146A. HUNECK, S. (1971). In *Progress in the Chemistry of Organic Natural Products*, **29**, Springer-Verlag, Wien.
147. HUNECK, S. and FOLLMANN, G. (1964). *Z. Naturf.*, **19**, 658.
148. HUNECK, S. and FOLLMANN, G. (1965). *Z. Naturf.*, **20**, 496.
149. HUNECK, S. and SIEGEL, M. (1963). *Naturwissenschaften*, **50**, 154.
149A. JACOBS, J. B. and AHMADJIAN, V. (1971). *J. Phycol.*, **7**, 71.
150. JAMES, P. W. (1965). *Lichenologist*, **3**, 95.
151. JOHNSON, G. T. (1940). *Ann. Mo. bot. Gdn*, **27**, 1.
152. JOHNSON, G. T. (1954). *Mycologia*, **46**, 339.
153. JOHNSON, T. W. and SPARROW, F. K. (1961). *Fungi in Oceans and Estuaries*, J. Cramer, Weinheim.
154. JONES, E. W. (1952). *Revue bryol. lichen.*, **21**, 96.
155. JONES, J. M. (1966). Unpublished data.
156. KAJANUS, B. (1911). *Ark. Bot.*, **10** (4), 1.
157. KARI, L. E. (1936). *Turun Yliop. Julkaisu*, **5** (1), 1.
158. KERSHAW, K. A. (1961). *Lichenologist.* **1**, 251.
159. KERSHAW, K. A. (1964). *Lichenologist*, **2**, 263.
159A. KERSHAW, K. A. and MILLBANK, J. W. (1969). *Lichenologist*, **4**, 83.
159B. KERSHAW, K. A. and MILLBANK, J. W. (1970). *Lichenologist*, **4**, 214.
159C. KLEMENT, O. (1955). *Beih. Repert. Spec. nov. Regni veg.*, **135**, 1.
160. KORF, R. P. (1958). *Sci. Rep. Yokohama natn. Univ.*, **7**, 7.
161. KROG, H. (1951). *Nytt Mag. Naturvid.*, **88**, 57.
162. KUROKAWA, S. (1962). *Beih. Nova Hedwigia*, **6**, 1.
163. KUROKAWA, S. (1965). *J. Jap. Bot.*, **40**, 264.
163A. KUROKAWA, S. (1971). *J. Jap. Bot.*, **46**, 297.
164. KUROKAWA, S. and JINZENJI, Y. (1965). *Bull. natn. Sci. Mus., Tokyo*, **8**, 369.
165. LAMB, I. M. (1951). *Can. J. Bot.*, **29**, 522.
166. LAMB, I. M. (1954). *Rhodora*, **56**, 105.
167. LAMB, I. M. (1955). *Farlowia*, **4**, 423.
168. LAMB, I. M. (1956). *Lloydia*, **19**, 157.
169. LAMB, I. M. (1963). *Index Nominum Lichenum*, Ronald Press, New York.
170. LAMB, I. M. (1964). *Brit. Antarct. Surv. Scient. Rep.*, **38**, 1.
171. LAMB, I. M. (1966). Unpublished data.
172. LAMBINON, J., MAQUINAY, A. and RAMAUT, J. L. (1964). *Bull. Jard. bot. État Brux.*, **34**, 273.
173. LANGE, O. L. (1953). *Flora, Jena*, **140**, 39.
174. LANGE, O. L. (1957). *Natur Volk*, **87**, 266.

175. LANGE, O. L. (1963). *Ber. dt. bot. Ges.*, **75**, 351.
176. LANGE, O. L. and ZIEGLER, H. (1963). *Mitt. flor.-soz. ArbGemein. N.F.*, **10**, 156.
176A. LAUNDON, J. R. (1957). *Lond. Nat.*, **37**, 66.
176B. LAUNDON, J. R. (1967). *Lichenologist*, **3**, 277.
176C. LEBLANC, F. (1969). In *Proceedings of the First European Congress on the Influence of Air Pollution on Plants and Animals*, Wageningen.
177. LEROY, L. W. and KOKSOY, M. (1962). *Econ. Geol.*, **57**, 107.
178. LEWIN, R. A. (1962). *Physiology and Biochemistry of Algae*, Academic Press, New York.
179. LINDAHL, P. O. (1960). *Svensk bot. Tidskr.*, **54**, 565.
180. LINDBERG, B., MISIORNY, A. and WACHTMEISTER, C. A. (1953). *Acta chem. scand.*, **7**, 591.
181. LINKO, P., ALFTHAN, M., MIETTINEN, J. K. and VIRTANEN, A. J. (1953). *Acta chem. scand.*, **7**, 1310.
182. LINKOLA, K. (1918). *Meddn Soc. Fauna Flora fenn.*, **44**, 153.
183. LLANO, G. A. (1944). *Bot. Rev.*, **10**, 1.
184. LLANO, G. A. (1956). *Econ. Bot.*, **10**, 367.
185. LOOMAN, J. (1964). *Ecology*, **45**, 481.
186. LOUNAMAA, J. (1956). *Ann. Bot. Soc. Zool. Bot. Fenn.*, **39**, 1.
187. LOWE, J. L. (1939). *Lloydia*, **2**, 225.
188. LUTTRELL, E. S. (1955). *Mycologia*, **47**, 511.
189. MAASS, W. S. G., TOWERS, G. H. N. and NEISH, A. C. (1964). *Ber. dt. bot. Gesell.*, **77**, 157.
190. MAGNUSSON, A. H. (1934). *Acta Horti gothoburg.*, **9**, 43.
191. MAGNUSSON, A. H. (1939). *K. svensk Vet. Akad. Handl.*, **17** (5), 1.
192. MAGNUSSON, A. H. (1952). *Svensk bot. Tidskr.*, **46**, 178.
193. MELLOR, E. (1923). *Nature*, **112**, 299.
194. MĚRKA, V. (1951). *Publs Fac. Sci. Univ. Masaryk*, **4**, 97.
195. MIĆOVIĆ, V. M. and STEFANOVIĆ, V. D. (1961). *Bull. Acad. serbe Sci. Arts*, **26**, 113.
195A. MILLBANK, J. W. and KERSHAW, K. A. (1969). *New Phytol.*, **68**, 721.
196. MITCHELL, J. C. and CHAMPION, R. H. (1965). *Bryologist*, **68**, 116.
197. MITTAL, O. P. and SESHADRI, T. R. (1954). *J. scient. ind. Res.*, **13A**, 174.
198. MÖLLER, A. (1887). *Unters. Bot. Inst. Kön. Akad. Munster*, **1**, 1.
199. MOORE, R. T. (1960). *Mycologia*, **52**, 805.
200. MOREAU, F. (1928). *Les Lichens*, Lechevalier, Paris.
201. MOSBACH, K. (1964). *Acta chem. scand.* **18**, 329.
202. MURTY, T. K. and SUBRAMANIAN, S. S. (1958). *J. scient. ind. Res.*, **17C**, 105.
203. MUSSO, H. (1960). *Planta med.*, **8**, 432.
204. NANNFELDT, J. A. (1932). *Nova Acta R. Soc. Scient. upsal.*, ser. 4, **8** (2), 1.
205. NEELAKANTAN, S. (1965). 'Recent Developments in the chemistry of lichen substances', in *Advancing Frontiers in the Chemistry of Natural Products*, Hindustani Publishing Corp., Delhi.
206. NIENBURG, W. (1919). *Z. Bot.*, **11**, 1.
207. NORDIN, I. (1964). *Svensk bot. Tidskr.*, **58**, 225.
208. NYLANDER, W. (1866). *Flora*, **49**, 198.
208A. OBERWINKLER, F. (1970). *Dt. bot. Gesell. Neue Folge*, **4**, 139.
209. OZENDA, P. (1963). 'Lichens' in *Encyclopedia of Plant Anatomy*, ZIMMERMAN, W. and OZENDA, P. Borntraeger, Berlin.

210. PALMER, H. E., HANSON, W. C., GRIFFIN, B. I. and ROESCH, W. C. (1963). *Science, N.Y.,* **142,** 64.
211. PAULSON, R. (1921). *Trans. Br. mycol. Soc.,* **7,** 41.
212. PEARSON, L. and SKYE, E. (1965). *Science, N.Y.,* **148,** 1600.
212A. PEVELING, E. and VAHL, J. (1968). *Beitr. Elektronenmikrosk. Direkt-abbild. Oberflachen,* **1,** 205.
213. PHILLIPS, H. C. (1963). *J. Tenn. Acad. Sci.,* **38,** 95.
214. PLESSL, A. (1963). *Öst. bot. Z.,* **110,** 194.
215. PLITT, C. C. (1934). *Bryologist,* **37,** 102.
216. POELT, J. (1958). *Planta,* **52,** 600.
216A. POELT, J. (1972). *Bot. Notiser,* **125,** 77.
216B. POELT, J. and JÜLICH, W. (1969). *Herzogia,* **1,** 331.
217. PORTER, L. (1917). *Proc. R. Ir. Acad.,* **34,** 17.
217A. PYATT, B. F. (1970). *Environ. Pollution,* **1,** 45.
218. QUISPEL, A. (1945). *Recl Trav. bot. néerland.,* **40,** 413.
218A. RANWELL, D. S. (1968). *Lichenologist,* **4,** 55.
219. RAO, D. N. and LEBLANC, F. (1965). *Bryologist,* **68,** 284.
220. RAO, D. N. and LEBLANC, F. (1966). *Bryologist,* **69,** 69.
220A. RAO, D. N. and LEBLANC, F. (1967). *Bryologist,* **70,** 141.
221. RAUP, L. C. (1930). *Bryologist,* **33,** 57.
222. REINKE, J. (1896). *Jb. wiss. Bot.,* **29,** 171.
222A. RICHARDSON, D. H. S. (1970). *Lichenologist,* **4,** 350.
223. RICHARDSON, D. H. S. and MORGAN-JONES, G. (1964). *Lichenologist,* **2,** 205.
224. RIED, A. (1960). *Flora,* **149,** 345.
224A. RIED, A. (1960). *Flora,* **148,** 612.
225. ROBINSON, H. R. (1959). *Bryologist,* **62,** 254.
226. RUNEMARK, H. (1956). *Op. bot. Soc. bot. Lund.,* **2,** 1.
227. RYDZAK, J. (1955). *Annls Univ. Mariae Curie-Skłodowska,* **10,** 321.
228. RYDZAK, J. (1957). *Annls Univ. Mariae Curie-Skłodowska,* **10,** 87.
229. RYDZAK, J. (1961). *Annls Univ. Mariae Curie-Skłodowska,* **16,** 1.
230. SALISBURY, F. B. (1962). *Science, N.Y.,* **136,** 17.
230A. SANTESSON, J. (1969). *Acta univ. Upsal.,* **127,** 7.
230B. SANTESSON, J. (1969). *Arkiv Kemi,* **30,** 363.
231. SANTESSON, R. (1939). *Meddn Lunds Univ. Limnol. Inst.,* **1,** 1.
232. SANTESSON, R. (1952). *Symb. bot. upsal.,* **12,** 1.
233. SATO, M. (1965). *Bryologist,* **68,** 320.
234. SAVICZ, V. P., LITVINOV, M. A. and MOISSEJEVA, E. N. (1960). *Planta med.,* **8,** 191.
235. SCHATZ, A., CHERONIS, N. D., SCHATZ, V. and TRELAWNY, G. S. (1954). *Proc. Pa. Acad. Sci.,* **30,** 62.
236. SCHMIDT, A. (1953). *Mem. Mus. Nat. Hist. Nat.* ser. B., *N.S.,* **3,** 1.
237. SCHOLANDER, P. F., FLAGG, W., WALTERS, V. and IRVING, L. (1952). *Am. J. Bot.,* **39,** 707.
238. SCHWENDENER, S. (1867). *Verh. schweiz. naturf. Ges. Rheinfelden,* **1867,** 88.
239. SCOTT, G. D. (1956). *New Phytol.,* **55,** 111.
240. SCOTT, G. D. (1959). *Lichenologist,* **1,** 109.
241. SCOTTER, G. W. (1963). *Can. J. Bot.,* **41,** 1199.
242. SCOTTER, G. W. (1965). *Can. J. Pl. Sci.,* **45,** 246.
243. SHIBATA, S. (1964). 'Biogenetical and Chemotaxonomical Aspects of Lichen Substances', in *Beiträge zur Biochemie und Physiologie von Naturstoffen.* 451. Gustav Fischer, Jena.

244. SHIBATA, S. and HSÜCH-CHING CHIANG (1965). *Phytochem.*, **4**, 133.
245. SHIBATA, S., NATORI, S. and UDAGAWA, S. (1964). *List of Fungal Products*, University of California Press, Berkeley.
246. SJÖSTRÖM, A. G. M. and ERICSON, L. E. (1953). *Acta chem. scand.*, **7**, 870.
247. SKYE, E. (1958). *Svensk bot. Tidskr.*, **52**, 133.
247A. SKYE, E. and HALLBERG, I. (1969). *Oikos*, **20**, 547.
248. SMITH, A. L. (1921). *Lichens*, Cambridge University Press, London.
249. SMITH, D. C. (1961). *Lichenologist*, **1**, 209.
250. SMITH, D. C. (1962). *Biol. Rev.*, **37**, 537.
251. SMITH, D. C. (1963). *Symp. Soc. gen. Microbiol.*, **13**, 31.
252. SMITH, D. C. (1966). Unpublished data.
253. SMITH, D. C. and DREW, E. A. (1965). *New Phytol.*, **64**, 195.
253A. SMITH, P. M. (in preparation). *An Introduction to Chemotaxonomy of Plants*, 'Contemporary Biology Series', Edward Arnold, London.
254. SMYTH, E. S. (1934). *Ann. Bot.*, **48**, 781.
255. SOLBERG, Y. J. (1956). *Acta chem. scand.*, **10**, 1116.
255A. SOWTER, F. A. (1971). *Lichenologist*, **5**, 176.
256. STEVENS, R. B. (1941). *Am. J. Bot.*, **28**, 59.
257. STOLL, A., BRACK, A. and RENZ, J. (1947). *Experientia*, **3**, 115.
258. SUBBOTINA, E. N. and TOMOFÉEFF-RESSOVSKY, N. V. (1961). *Bot. Zh. S.S.S.R.*, **46**, 212.
259. SUBRAMANIAN, S. S. and RAMAKRISHNAN, S. (1964). *Curr. Sci.*, **33**, 522.
260. SWINSCOW, T. D. V. (1960). *Lichenologist*, **1**, 169.
261. TARGÉ, A. and LAMBINON, J. (1965). *Bull. Soc. r. Bot. Belg.*, **98**, 295.
262. THOMAS, E. A. (1939). *Beitr. KryptogFlora Schweiz*, **9**, 1.
263. THOMSON, J. W. (1948). *Bull. Torrey bot. Club*, **75**, 486.
264. THOMSON, J. W. (1963). *Beih. Nova Hedwigia*, **7**, 1.
264A. TIBELL, L. (1971). *Svensk bot. Tidskr.*, **65**, 138.
265. TOBLER, F. (1911). *Jb. wiss. Bot.*, **49**, 389.
266. TOBLER, F. (1925). *Biologie der Flechten*, Borntraeger, Berlin.
267. TOMASELLI, R. (1950). *Arch. Bot. Biogr. Ital.*, **26**, 1.
268. TOMASELLI, R. (1957). *Arch. Bot. Biogr. Ital.*, **33**, 1.
269. TOMASELLI, R. (1959). *Atti Ist. bot. Univ. Lab. crittogam. Pavia*, **16**, 180.
270. TOMASELLI, R. (1962). *Atti Accad. gioenia Sci. Nat.*, **16**, 168.
271. UYENCO, F. (1963). *Bryologist*, **66**, 217.
272. VAINIO, E. A. (1890). *Acta Soc. Fauna Flora fenn.*, **7** (7) ,1.
273. VOIGT, G. K. (1960). *Am. Midl. Nat.*, **63**, 321.
274. WACHTMEISTER, C. A. (1956). *Bot. Notiser*, **109**, 313.
275. WACHTMEISTER, C. A. (1958). *Svensk kem. Tidskr.*, **70**, 117.
276. WACHTMEISTER, C. A. (1958). *Acta chem. scand.*, **12**, 147.
277. WARD, H. M. (1884). *Trans. Linn. Soc. Lond.*, **2**, 87.
278. WATSON, W. (1919). *J. Ecol.*, **7**, 71.
279. WATSON, W. (1953). *Census Catalogue of British Lichens*, Cambridge University Press, London.
280. WEBER, W. A. (1965). *Svensk bot. Tidskr.*, **59**, 59.
281. WERNER, R. G. (1930). *Bull. Soc. mycol. Fr.*, **46**, 199.
282. WERNER, R. G. (1931). *Mém. Soc. Sci. nat. phys. Maroc*, **27**, 7.
282A. WETHERBEE, R. (1969). *Michigan Bot.*, **8**, 170.
283. WILHELMSEN, J. B. (1959). *Bot. Tidskr.*, **55**, 30.
284. WIRTH, M. and HALE, M. E. (1963). *Contr. U.S. natn. Herb.*, **36**, 63.

285. ZAHLBRUCKNER, A. (1926). 'Lichenes (Flechten)' in *Die Natürlichen Pflanzenfamilien*, ENGLER, A. and PRANTL, K. Vol. 8, 2nd edn., Engelmann, Leipzig.
286. ZAHLBRUCKNER, A. (1922). *Catalogus Lichenum Universalis*, vol. 1–9, Gebrüder Borntraeger, Leipzig (Johnson Reprint Corp., New York).
287. ZEHNDER, A. (1949). *Ber. schweiz. bot. Ges.*, **59**, 201.
288. ZOPF, W. (1907). *Die Flechtenstoffe in chemischer, pharmakologischer, und technischer Beziehung*, Jena.
289. ZOPF, W. (1913). *Beih. bot. Zbl.*, **14**, 95.

Index